Carbohydrates 1

1   Disaccharides can be split by:

☐ **A** hydrolysis of glycosidic bonds

☐ **B** condensation of glycosidic bonds

☐ **C** hydrolysis of ester bonds

☐ **D** condensation of ester bonds   **(1 mark)**

2   The diagram shows an α-glucose molecule.

Draw a diagram to show the products formed when two α-glucose molecules join together.

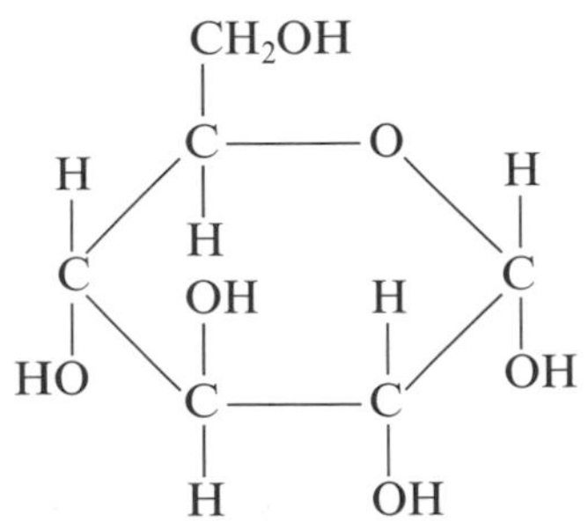

(3 marks)

3   Complete the following table about disaccharides.

Disaccharide	Component monosaccharides	Type of bond between monosaccharides
sucrose		
lactose		β-1,4 glycosidic
maltose	2 α-glucose	

(3 marks)

4   Compare and contrast the structures of α- and β-glucose.

> You must discuss both similarities and differences in compare and contrast questions and look at both angles.

....

....

....

....   **(2 marks)**

Carbohydrates 2

Guided 1 Both glycogen and starch are energy storage compounds in living organisms. Explain the features of the two molecules that adapt them to this role.

Both glycogen and starch are made from glucose molecules which

....

....

These bonds are easily broken by ....

....

....

.... **(4 marks)**

2 The table shows some statements about polysaccharides. Indicate with a tick (✓) which of the following statements are correct for cellulose.

Statement	Correct
polymer of glucose	
molecule contains α- and β-glucose	
glycosidic bonds present	
molecule may have side branches	
molecule can form hydrogen bonds with adjacent molecules	

(5 marks)

3 Starch is a mixture of amylose and amylopectin. Explain why this allows it to provide both a quick and a slow release of glucose and thus energy after it is eaten.

....

....

....

.... **(2 marks)**

4 Describe the structure of a cellulose fibre.

> For 'describe questions', you do not need to make a judgement or explain how or why, you only need to describe what is requested.

....

....

....

....

.... **(3 marks)**

Lipids

1   The diagram shows part of the structural formula of a monounsaturated fatty acid with the formula $C_{10}H_{21}COOH$. Complete the diagram of the molecule.

> Note that a structural formula always shows all the bonds, so just writing in the missing part is not enough.

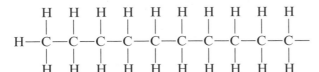

(2 marks)

2   The hydrocarbon chain shown in the diagram is saturated. Redraw it to show an unsaturated chain with the same number of carbon atoms.

> Think about what 'unsaturated' means and make sure each carbon has four bonds.

(2 marks)

3   The table contains some statements about triglycerides. Indicate with a tick (✓) which of the following statements is correct.

Statement	Correct
Triglycerides are the building blocks of proteins.	
Triglycerides can be modified to form phospholipids.	
Water is released when a triglyceride molecule is hydrolysed.	
Some triglycerides contain sulfur.	

(1 mark)

4   Draw a triglyceride using the shapes shown. You may use each one once, more than once, or not at all.

☐ glycerol   ∿∿∿ fatty acid   ——— ester bond

(3 marks)

3

Functions of lipids

1   Draw a labelled, annotated diagram to show the structure of a phospholipid molecule in relation to its role as part of the membrane.

> 'Annotated' means that you should label the diagram with the names of its parts and the relevant important features of those parts.

(5 marks)

Guided

2   Explain the features of lipid molecules which adapt them for their roles in energy storage and insulation.

> 'Explain' tells you that you need to describe something and then why. The word 'because' is very useful in an answer to this kind of question.

Lipids are good for energy storage because they yield .... × as

much energy as both .... and .... This is because

....

....

....

....

....   (4 marks)

3   Explain why phospholipids form bilayers when they are mixed with water.

....

....

....

....

....   (3 marks)

4   State two ways in which a membrane may become more fluid.

....

....

....

....   (2 marks)

Proteins: amino acids and polypeptides

1   The bond between two amino acids is:

    ☐ **A**   a glycosidic bond

    ☐ **B**   a peptide bond

    ☐ **C**   a phosphodiester bond

    ☐ **D**   an ester bond   **(1 mark)**

2   (a)   The diagram shows parts of two amino acids. Complete it to show them joined together.

                                                              (2 marks)

    (b)   State the name of the molecule you have drawn.

      ....   **(1 mark)**

> **Guided**

    (c)   Are the two amino acids the same? Explain your answer.

      They are not the same because ....

      ....

      ....

      ....   **(2 marks)**

3   The diagram shows an amino acid. Name the part, A, B, C or D, that makes all amino acids acidic.

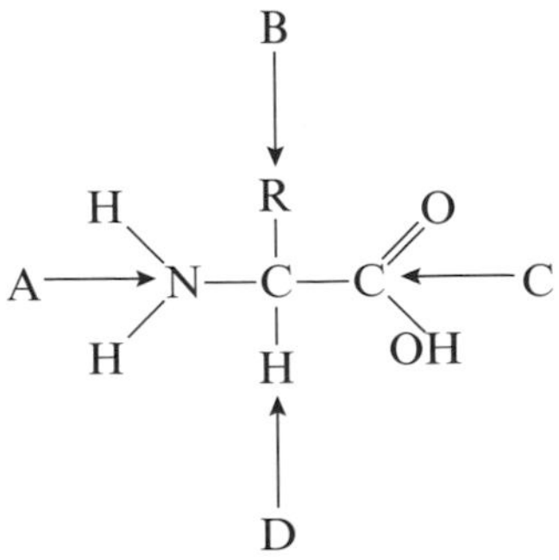

     Answer ....   **(1 mark)**

4   Which of the following describes a chain of amino acids?

    ☐ **A**   primary structure of a protein

    ☐ **B**   secondary structure of a protein

    ☐ **C**   tertiary structure of a protein

    ☐ **D**   quaternary structure of a protein   **(1 mark)**

Protein structures

The diagram shows an enzyme's primary, secondary and tertiary structure.

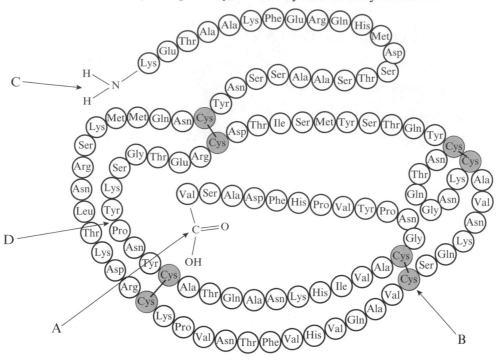

> **Guided**

1   *With reference to the diagram, explain what is meant by the terms primary structure, secondary structure and tertiary structure.

> A star (*) next to a question means marks will be awarded for your ability to structure your answer logically, showing how the points that you make are related to or follow on from each other where appropriate. This question asks for reference to the diagram so you must quote from it in your answer.

Primary structure is the sequence of ....

....

....

Secondary structure is the part of the structure stabilised by ....

....

....

Tertiary structure is the ....

....

....   **(6 marks)**

2   State the names of the groups labelled A and C.

A is ....   C is ....   **(2 marks)**

Haemoglobin and collagen

> **Guided**

1 Contrast fibrous proteins with globular proteins.

> Make it clear what the difference is by writing your answer as a comparison. An example is given.

Fibrous proteins are straight chains whereas globular are spherical. ....

....

....

....

.... **(3 marks)**

2 Describe the structure of collagen.

> For 'describe' questions, you do not need to make a judgement or explain how or why, you only need to describe what is requested.

> This answer requires extended prose and will need some planning. Jot down five facts you know about the structure of collagen then assemble them into prose.

....

....

....

....

....

....

.... **(5 marks)**

> **Guided**

3 Haemoglobin is a globular protein with four haem groups. It undergoes a shape change when it binds with a molecule of oxygen. Explain how these features adapt it to its function of picking up oxygen.

Haemoglobin is globular, so it is soluble in water. ....

....

....

.... **(3 marks)**

4 Explain how the quaternary structure of haemoglobin leads to cooperative binding.

....

....

....

.... **(3 marks)**

Nucleic acids: DNA

1   DNA consists of a double helix composed of two strands held together by:

    ☐ **A**   hydrogen bonds between the bases of nucleotides

    ☐ **B**   phosphodiester bonds between the bases of nucleotides

    ☐ **C**   hydrogen bonds between the R groups of amino acids

    ☐ **D**   phosphodiester bonds between the R groups of amino acids   **(1 mark)**

> Think about two ideas for this question – what DNA is made of and what types of bonds are used.

2   The following are simple diagrams of deoxyribose, phosphate and a base. Using these, draw a nucleotide in the space below.

> You must use the shapes given and the first step is to identify which shape relates to which unit, e.g. deoxyribose is a pentose sugar. You then need to think about what is joined to what and where the bonds are.

(3 marks)

Guided   3   Describe the diagram you have drawn above.

The phosphate group is attached to the deoxyribose sugar at carbon atom

number five by a ....

....

....

.... **(3 marks)**

4   State the base pairing rules in DNA.

....

....

.... **(2 marks)**

5   Explain why a purine cannot pair with a purine and why a pyrimidine cannot pair with a pyrimidine.

....

....

.... **(2 marks)**

Nucleic acids: RNA

1   Compare and contrast the role of hydrogen bonds in the structure of DNA and transfer RNA (tRNA).

> You must mention at least one similarity and one difference.

....

....

....

....

....

....

(3 marks)

2   Which of the following pairs of features is found in tRNA?

☐ **A**   an anticodon and an amino acid binding site

☐ **B**   an anticodon and a nucleotide binding site

☐ **C**   a codon and an amino acid binding site

☐ **D**   a codon and a nucleotide binding site

(1 mark)

3   Which of the following do mRNA and tRNA have in common?

☐ **A**   base pairs

☐ **B**   sugar phosphate backbone

☐ **C**   codons

☐ **D**   anticodons

(1 mark)

4   Which of the following is true of both mRNA and tRNA compared to DNA?

☐ **A**   mRNA and tRNA have thymine instead of uracil which is found in DNA

☐ **B**   mRNA and tRNA have uracil instead of adenine which is found in DNA

☐ **C**   mRNA and tRNA have uracil instead of thymine which is found in DNA

☐ **D**   mRNA and tRNA have adenine instead of uracil which is found in DNA

(1 mark)

5   (a)   Name the molecule shown in the diagram.

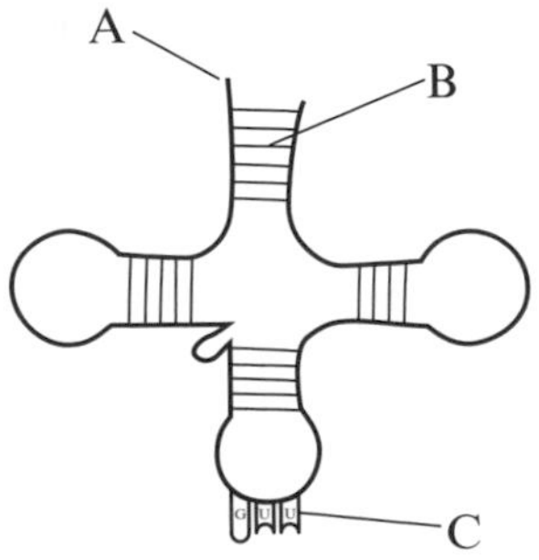

....

(1 mark)

(b)   Identify the parts labelled A, B and C.

A is ....

B is ....

C is ....

(3 marks)

Had a go ☐ Nearly there ☐ Nailed it! ☐

DNA replication and the genetic code

1 Compare and contrast a gene and a codon.

> Make sure you link the characteristics of the gene with the same characteristics for the codon and give at least one similarity and one difference.

....

....

....

....

....

.... **(3 marks)**

Guided

2 Explain **one** advantage of degeneracy in the genetic code.

> In this case, 'explain' means that you will need to describe what degeneracy is first. Also notice the bold print; this means only one advantage will gain any credit.

Degeneracy is where amino acids are ....

....

....

....

....

.... **(2 marks)**

3 What is the correct order of the four steps in DNA replication shown below?

A The enzymes DNA polymerase and DNA ligase join the nucleotides together.

B Hydrogen bonding links the two strands together.

C The two strands of DNA unwind and split apart.

D Free nucleotides line up along each strand, observing the complementary base pairing rules.

Step 1 is .... Step 3 is ....

Step 2 is .... Step 4 is .... **(2 marks)**

4 If an organism was supplied with nucleotides containing a heavy isotope and allowed to replicate once using these, what would be the mass of the new DNA formed?

☐ **A** normal

☐ **B** heavy

☐ **C** between normal and heavy

☐ **D** no DNA could be made with this isotope **(1 mark)**

Protein synthesis

1 (a) Name the substances X, Y and Z in the diagram.

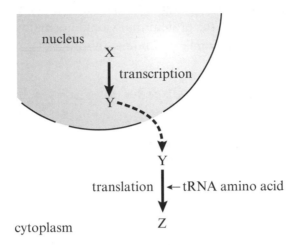

X is ....

Y is ....

Z is ....   **(3 marks)**

> **Guided**

  (b) Describe the processes of transcription and translation.

Transcription is when one of the strands of DNA is used as a ....

....

....

Translation is when a protein is synthesised ....

....

....   **(5 marks)**

2 Describe what might happen to a polypeptide to make it into a functional protein, such as an enzyme.

> In your answer you need to distinguish between a polypeptide and a protein and think about what is important for a protein to be functional.

....

....

....

....   **(2 marks)**

Mutation

Guided ⟩

1 Krabbe disease is caused by mutations in a gene called GALC. This leads to a lack of the enzyme galactocerebrosidase. Explain how a mutation in the GALC gene could result in a lack of the protein that would normally become the active enzyme galactocerebrosidase, and explain how this might stop the formation of an active enzyme.

A gene is a sequence of bases that codes for a sequence of amino acids in a protein.

A mutation is a change in that sequence, so ....

....

....

....

....

(4 marks)

2 The CFTR protein is either not synthesised or has reduced or lack of functionality in a condition called cystic fibrosis. How might a mutation lead to

(a) no synthesis of CFTR?

....

....

....

....

(2 marks)

(b) synthesis of CFTR but with reduced functionality?

....

....

....

....

(2 marks)

Sickle cell anaemia

1 A blood disorder called sickle cell anaemia is caused by a mutation of the gene which codes for haemoglobin. The diagram shows the DNA, mRNA and resultant amino acids for the relevant section of the normal gene and the mutated one.

DNA: C-A-C-C-T-G-G-A-C-T-G-A-G-G-G-G-A-C-T-C-C-T-C
RNA: G-U-C-G-A-C-C-U-G-A-C-U-C-C-U-G-A-G-G-A-G

<u>Val</u>   <u>His</u>   <u>Leu</u>   <u>Thr</u>   <u>Pro</u>   <u>Glu</u>   <u>Glu</u>
normal

DNA: C-A-C-C-T-G-G-A-C-T-G-A-G-G-G-G-A-C-A-C-C-T-C
RNA: G-U-C-G-A-C-C-U-G-A-C-U-C-C-U-G-U-G-G-A-G

<u>Val</u>   <u>His</u>   <u>Leu</u>   <u>Thr</u>   <u>Pro</u>   <u>Val</u>   <u>Glu</u>
mutated

What type of mutation is the example shown above?

☐ **A** insertion mutation

☐ **B** deletion mutation

☐ **C** translocation mutation

☐ **D** substitution mutation

> Look at the amino acid sequence and identify which has been changed then backtrack to the base sequence to see what the change is.

(1 mark)

Guided 2 Explain why the red blood cells change shape in sickle cell anaemia.

> In a question like this, you must try to explain what is asked for in a logical sequence rather than just writing out separate points from your knowledge.

The condition sickle cell anaemia is caused by a mutation in the gene for the beta globin portion of the haemoglobin molecule. ....

....

....

....

....

.... **(4 marks)**

Guided 3 Describe two symptoms of sickle cell anaemia and explain how they arise from the condition.

> If you are given a choice like this, choose two examples that you are sure you can explain.

Symptom 1: anaemia ....

Explanation: ....

....

Symptom 2: ....

Explanation: damage to the spleen by ....

.... **(4 marks)**

Enzymes

1   An amoeba is a small single-celled animal that engulfs bacteria for food. The bacteria
    are digested inside the cell by enzymes. These enzymes are examples of which type?

   ☐ **A** extracellular   ☐ **B** intercellular   ☐ **C** intracellular   ☐ **D** extercellular   **(1 mark)**

2   (a) Caffeine is converted in the liver by an enzyme complex
         called cytochrome P450 oxidase. Three products are
         formed as shown in the diagram. Explain how this
         shows that the cytochrome P450 oxidase complex must
         be more than one enzyme.

> Each of the products has the same general formula, $C_7H_8N_4O_2$, so you must think about how they differ.

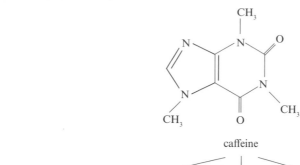

caffeine

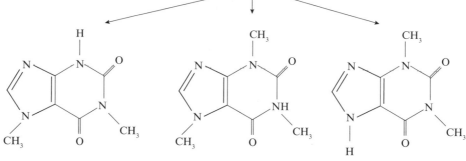

paraxanthine   theobromine   theophylline

....

....

....

....

....

....   **(3 marks)**

   (b) Paraxanthine : theobromine : theophylline are produced
        in the ratio 84 : 12 : 4. Assuming 50 milligrams of
        paraxanthine were produced after a cup of coffee,
        calculate the mass of theobromine and theophylline this person would produce.
        Show your working.

> Notice that 12 divides into 84 seven times.

        theobromine ....mg

        theophylline ....mg   **(3 marks)**

Activation energy and catalysts

1 Which of the following statements is true of enzymes?

☐ **A** They cause reactions to happen by raising the activation energy.

☐ **B** They increase the rate of reaction by lowering the activation energy.

☐ **C** They increase the rate of reaction by raising the activation energy.

☐ **D** They cause reactions to happen by lowering the activation energy.

> There are two parts to each answer. Decide for each part whether it is correct then see which response has both parts correct.

(1 mark)

2 Enzymes provide an alternative pathway for reactions at their active sites.
Explain what happens.

> This question is about what happens in the active site when the substrate or substrates are in there. To answer this fully, you will need to think about enzymes **both** breaking and making bonds.

....

....

....

....

....

.... **(3 marks)**

3 Look at the energy diagram of a reaction with and without an enzyme. Explain what A, B and C show.

A ....

....

B ....

....

C ....

.... **(3 marks)**

Reaction rates

Practical skills

1   A student was asked to design an experiment to look at the effect of pH on the initial rate of the enzyme reaction in which hydrogen peroxide is broken down into water and oxygen. She proposed a procedure in which she would time how long it took for $100 \, cm^3$ of oxygen to be produced at each of five pH values. The rate would then be calculated by finding the reciprocal of the time taken (1 / time). Explain what is wrong with her method for calculating the rate.

> Think about what happens to the amount of substrate as the reaction proceeds.

....

....

....

....

....

.... **(4 marks)**

2   The graphs show the results of two experiments to determine the initial rate of two different enzyme-catalysed reactions. In experiment 1, a protease enzyme breaks down a suspension of a protein which was initially cloudy. In experiment 2, glucose phosphate in solution is converted into starch using starch synthetase enzyme. In both cases, the course of the reaction over time is followed using a colorimeter. Explain why the graphs are different shapes.

> Think about the products of each experiment.

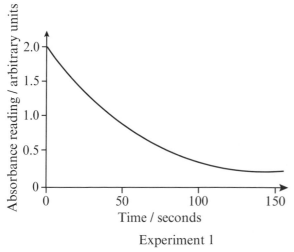

Experiment 1

Experiment 2

....

....

.... **(3 marks)**

Investigating enzyme activity

 Practical skills

1 Read this method for a practical designed to find the optimum temperature of an enzyme:

> Pipette 2 cm³ of 1% protein solution into a cuvette. Pipette 2 cm³ of 1% protease solution into the same cuvette. Mix thoroughly and immediately put into a colorimeter and start the stop clock. Measure and record the absorbance at suitable time intervals until there is little change. Repeat the procedure using a range of different temperatures ensuring that other conditions are unchanged.

(a) State the dependent variable in this experiment.

.... **(1 mark)**

(b) The method says 'ensuring that other conditions are unchanged'. Explain what this means and how it would be achieved with reference to this experiment.

> It is always important to read the stem of a question very carefully; here you need to do that to answer this question.

....

....

....

....

.... **(3 marks)**

Maths skills

(c) Explain how you would use the data to find the initial rate of the reaction for each temperature.

....

....

....

....

.... **(3 marks)**

2 State how temperature would be controlled in an experiment to look at the effect of pH on the same enzyme.

> You must be very precise, so 'in a water bath' is not sufficient detail.

....

....

....

.... **(2 marks)**

Factors affecting enzymes

1 Which of the four graphs, A, B, C or D, shows the effect of temperature on an enzyme-catalysed reaction?

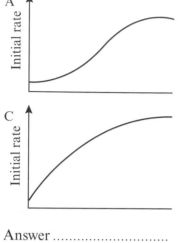

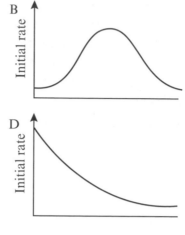

> You should know the shapes of the four enzyme factor curves, pH, temperature and enzyme and substrate concentration.

Answer ....

(1 mark)

2 The graph shows the effect of pH on an enzyme measured by two methods.
Analyse the data to explain the results shown by these two methods.

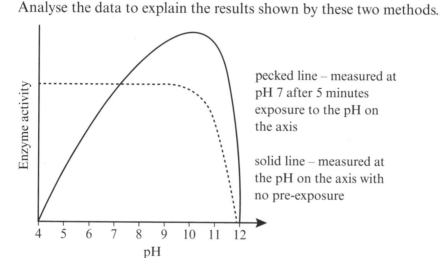

pecked line – measured at pH 7 after 5 minutes exposure to the pH on the axis

solid line – measured at the pH on the axis with no pre-exposure

> The word 'analyse' means that you need to comment on the data and then relate your comments to the situation being discussed to make a judgement.

> Notice the two lines intersect at pH 7 and that at greater pHs the one which did not receive any pre-exposure is always higher.

....

....

....

....

(3 marks)

3 Explain the effect of substrate concentration on the initial rate of an enzyme-catalysed reaction.

....

....

....

....

....

(4 marks)

Enzyme inhibition

1 (a) Succinic acid is converted into fumaric acid by the enzyme succinate dehydrogenase. Malonic acid affects the course of the reaction as shown in the graph. State what type of inhibitor malonic acid is and explain your answer.

> The malonic acid blocks the active site of the enzyme because it is similar in shape to succinic acid.

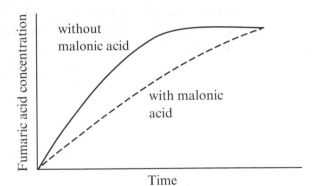

....

....

....

....

.... **(3 marks)**

(b) Draw the shapes of the curves you would expect with and without malonic acid on the axes below.

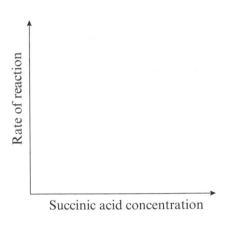

(2 marks)

2 An enzyme catalyses a reaction in which A is converted into B. A third substance, C, which does not resemble the structure of A, slows down the rate of the reaction. C is likely to be:

☐ **A** a competitive inhibitor

☐ **B** a non-competitive inhibitor

☐ **C** the end-product of the reaction

☐ **D** the substrate   **(1 mark)**

Water and ions

 Practical skills

1 *In hydroponics, soil is replaced with solutions containing the necessary minerals for plant growth. The mineral solutions are made up with the optimum concentration of each mineral, which has been determined by experiment. Design an experiment to determine the optimum concentration of calcium in the mineral solution used in the growth of a named plant.

> You need to be clear and logical in your answer so **plan** your answer **before** you start writing, e.g. think about the independent variable and the dependent variable and how you are going to change / control them. You should to try to support your answer with evidence and show that you have relevant knowledge. Read your answer through to make sure it is well structured, links together well and that your reasoning is clear.

....

....

....

....

....

....

....

....

.... **(6 marks)**

2 Complete the table below about the features of water that are important to living things.

Feature	Example of importance to living things
high specific heat capacity	
polar solvent	
surface tension	
incompressibility	
maximum density at 4 °C	

(5 marks)

Exam skills

1   (a)   The graph shows the effect of the enzyme lipase on
the initial rate of breakdown of lipid at different
concentrations. Analyse the data to explain the effect
of lipase on this reaction.

> The word
> 'analyse' means
> that you need
> to comment on
> the data and
> then relate your
> comments to
> the situation
> being discussed
> to make a
> judgement.

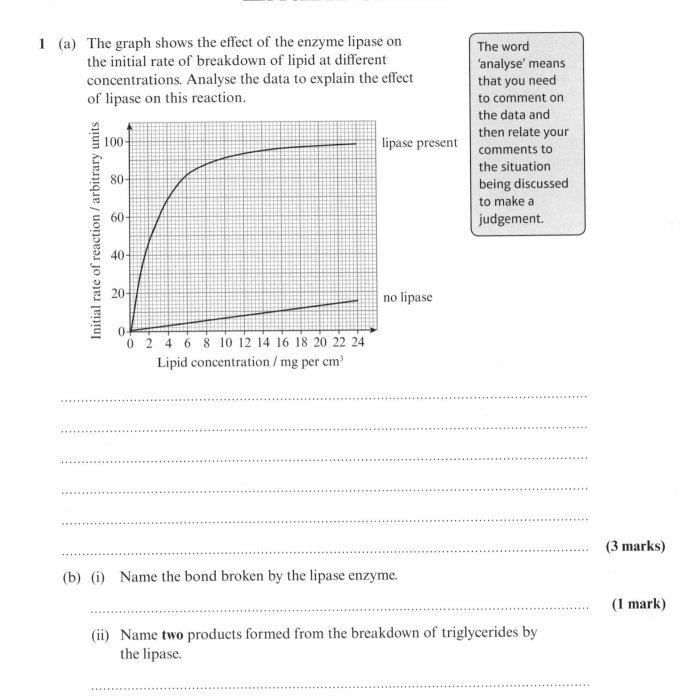

....

....

....

....

....

.... **(3 marks)**

   (b)   (i)   Name the bond broken by the lipase enzyme.

.... **(1 mark)**

   (ii)   Name **two** products formed from the breakdown of triglycerides by
the lipase.

....

.... **(2 marks)**

   (iii)   Explain how the breakdown of triglycerides would affect the pH of the
reaction mixture.

....

....

.... **(3 marks)**

Exam skills

1 The diagram shows the base sequence on a short section of DNA consisting of 12 mononucleotides. This base sequence contains the genetic code for a short section in the primary structure of a polypeptide.

Section of DNA	A	A	T	A	A	C	C	A	G	T	T	T

| Amino acids | leucine | leucine | valine | lysine |

(a) Name each of the bases represented by the letters, A, C, G and T, in the diagram.

A is ....

C is ....

G is ....

T is .... **(1 mark)**

(b) Using the sequence shown in the diagram, explain the meaning of each of the terms triplet, non-overlapping and degenerate code.

....

....

....

....

....

.... **(3 marks)**

(c) Which of the following are the names of two of the components, other than the bases, that form part of each mononucleotide in this sequence?

☐ A deoxyribose and nitrate   ☐ C ribose and nitrate

☐ B deoxyribose and phosphate   ☐ D ribose and phosphate **(1 mark)**

(d) Transcription of this section of DNA forms a complementary strand of mRNA. Describe how translation of this mRNA synthesises part of a polypeptide molecule.

> In questions of this kind where a context is provided, it is important to write your answer in terms of that context. In the case of (d), 'this mRNA' is the key phrase. Look for context in (b) as well.

....

....

....

....

....

....

....

.... **(5 marks)**

The cell theory

1 Look at the diagrams of various parts of animals and plants.

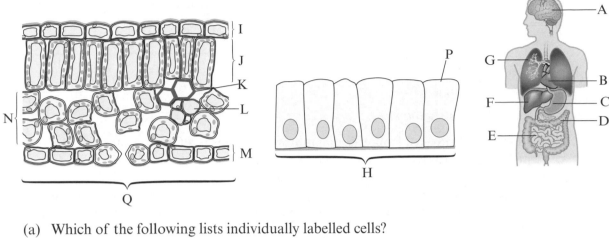

(a) Which of the following lists individually labelled cells?

☐ **A** K, L, H

☐ **B** P, L, M

☐ **C** J, K, L

☐ **D** K, L, P   **(1 mark)**

(b) State the four letters which label structures belonging to the same system and name the system.

....

.... **(2 marks)**

(c) Which of the following is an organ?

☐ **A** H

☐ **B** N

☐ **C** Q

☐ **D** J

> Look at the three diagrams and decide what each is representing then ask yourself which are cells, which are tissues and which are organs.

(1 mark)

> **Guided**

2 The cell theory states two things. One is that the cell is a fundamental unit of structure, function and organisation in all living organisms. What is the other?

New cells are ....

.... **(1 mark)**

Prokaryotes

1   The diagram shows a prokaryotic cell.

Note there are five marks available but only four label lines so one label line must require two parts.

(a) Label the parts shown.   **(5 marks)**

(b) Is it possible to tell from this diagram whether this is a Gram positive or a Gram negative bacterium? Explain your answer.

You could answer yes or no here, but you should choose the one for which you can give the best explanation since the marks will be for your explanation not for yes or no.

....

....   **(2 marks)**

2   Look at the table below showing some types of antibiotics and how they work.

Antibiotic and features	
A	Penicillins and cephalosporins inhibit the formation of the peptidoglycan layer of the bacterial cell wall.
B	Polymixins interact with the phospholipids of the outer membrane.
C	Other antibiotics target common processes such as protein synthesis by the ribosomes.

For each type of bacterium below, write the letter of the type of antibiotic which would be best used against it.

Both bacterial types have a peptidoglycan layer but Gram negative bacteria have the extra protection of the phospholipid-containing layer.

Gram negative ....

Gram positive ....

Both types ....   **(2 marks)**

> **Guided**   3   Describe two ways in which bacteria may benefit from having plasmids.

For 'describe' questions, you do not need to make a judgement or explain how or why, you only need to describe what is requested.

They contain genes that ....

....

....

....   **(3 marks)**

Eukaryotes

> **Guided**

1 A student wrote a description of each of the organelles seen in an electron micrograph image of an animal cell. Complete the student's table with the name of the organelle and a brief statement of its function.

Description	Name	Function
a large organelle with a double envelope with pores through it		stores DNA
a branching series of channels studded with small, roughly spherical, structures		
quite large oval organelles with folded membranes inside		
a pair of cylindrical structures at right angles to each other		makes the spindle fibre in cell division

(8 marks)

2 Complete the table below for the three cell types using the list below the table.

Not all the boxes need to have an entry.

Animal cell only	Plant cell only	Animal and plant	Bacteria only	All three cell types

(5 marks)

mitochondria   ribosomes   smooth endoplasmic reticulum (SER)

DNA in a nucleus   centriole   cell surface membrane

> **Guided**

3 State the organelles found in plant cells but not in animal cells and describe their functions.

Plant cells contain some organelles not found in animals. These include ....

.... where photosynthesis occurs, a vacuole where

.... and

.... are stored. Outside the plant cell

surface membrane is a ....   **(4 marks)**

Microscopy

Guided 1 State an advantage and a disadvantage of using an electron and a light microscope.

> The advantage of one will be the disadvantage of the other and vice versa.

Electron – Advantage: ....

....

Disadvantage: ....

....

Light – Advantage: ....

....

Disadvantage: ....

.... **(4 marks)**

2 A sample of blood viewed through a light microscope does not show evidence of bacteria in the blood even though the person is known to have a bacterial infection. Give two reasons why this would be the case.

> This is about visibility.

....

....

.... **(2 marks)**

Maths skills 3 Look at this photomicrograph of the surface of a plant leaf showing stomata. Use the magnification shown to calculate the actual length of a single guard cell.

> You will need to convert your measurement into micrometres and then think about what to do with the magnification factor.

×100

Answer .... **(4 marks)**

Had a go ☐ Nearly there ☐ Nailed it! ☐

Practical microscopy

1 The photographs show thin sections of the stem of *Arabidopsis thaliana*. The sections were stained with phloroglucinol-HCl, which stains lignin red.

| A | B | C |

Sectioned at: top   middle   base of stem
Magnification ×52

Key:
co, cortex;
f, fibre;
pi, pith;
x, xylem

Guided

(a) Comment on the distribution of lignified cells in this plant stem with reference to the tissues identified in the key.

Lignin is only seen in xylem cells at the top part of the plant but in the ....

....

....

.... **(3 marks)**

Maths skills

(b) Calculate the increase in actual diameter from the top to the base of the stem.

> You need to work out the actual size of the top and bottom of the stem on paper first.

Increase in diameter is .... **(3 marks)**

Use of the light microscope

1 The diagram shows two stages in the process to determine the size of cells and cell parts using the light microscope. Every small division on A is 0.01 mm long.

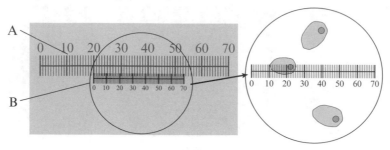

(a) Name the pieces of equipment A and B.

A is ....

B is ....   **(2 marks)**

(b) The part of the diagram on the right shows three cheek cells through the microscope. Determine how many times longer the cell is than its nucleus.

.... times   **(2 marks)**

(c) Calculate the length of a cheek cell.

> First calculate what one eyepiece micrometer unit is by knowing that 21 divisions equals 0.1 mm.

Length is ....   **(2 marks)**

2 The photograph shows some ciliated columnar epithelium from the uterus of a mammal. Draw and label three cells.

> Use single strong lines clearly showing outlines of cells etc.

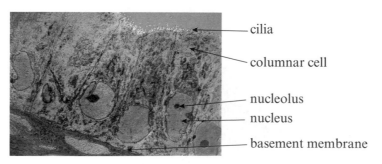

— cilia

— columnar cell

— nucleolus

— nucleus

— basement membrane

(3 marks)

Viruses: classification

> **Guided**

1 Viruses can be classified on the basis of their genetic material. This can be of two types, DNA or RNA. Explain, with examples, how there are **three** types of viruses.

> This is an 'explain' question, so it is not enough to state your point – you must also give reasons or say why you have come to that conclusion.

Because there are two types of RNA viruses ....

....

....

....

DNA viruses have ....

.... **(4 marks)**

2 The diagrams show three different types of viruses.

A   B   C

Match the letters on the diagrams to the virus types below.

DNA virus is ....   RNA virus is ....

Retrovirus is ....   **(2 marks)**

3 Which of the following lists of features is possessed by **all** viruses?

☐ **A** DNA, RNA

☐ **B** protein, 70S ribosomes

☐ **C** nucleic acid, protein

☐ **D** DNA, 70S ribosomes   **(1 mark)**

4 State which **one** of the seven characteristics of life viruses could be said to show and say why some scientists would disagree that they even show this.

> You will need to know the seven characteristics from GCSE and understand that viruses are usually classified as non-living.

....

....

.... **(2 marks)**

Viruses: replication

1   Norovirus is an RNA virus which causes stomach flu. It causes symptoms about 24 hours after eating the contaminated food, sometimes from as few as 20 viral particles. Explain how new viral particles are formed inside the host cells and in sufficient quantities to cause symptoms after 24 hours, starting from 20 particles.

> Do not be confused by norovirus being chosen for this question; the question is simply about the rapid replication of viruses.

....

....

....

....

....

....

.... **(5 marks)**

2   The diagram shows the cycle of the retrovirus HIV. Match the letters on the diagram with the following actions of anti-viral drugs.

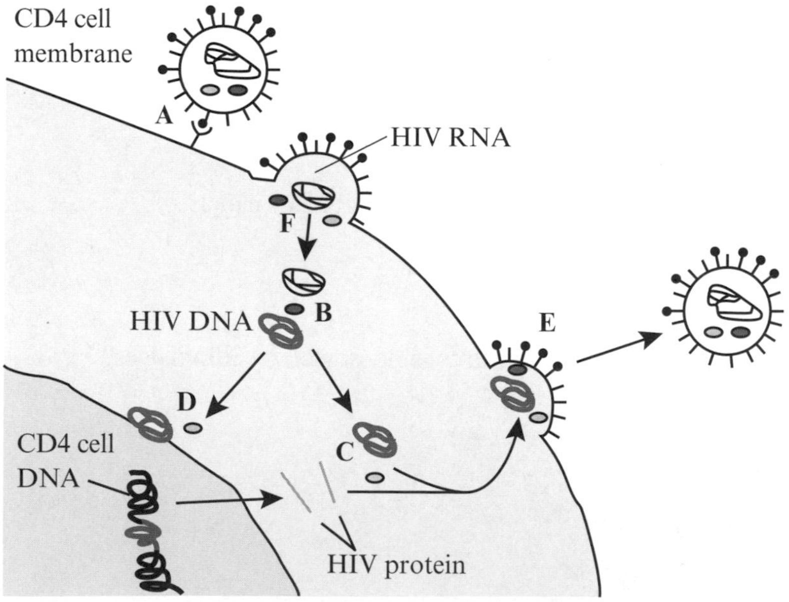

CD4 cell membrane

HIV RNA

A

F

HIV DNA   B   E

D

CD4 cell DNA   C

HIV protein

> There are six letters but only four actions so two letters are not relevant. Decide what is happening at each stage in the diagram then find the description of it in the actions.

Target the receptors by which viruses recognise their host cells ....

Target integrase ....

Target protease enzymes ....

Target reverse transcriptase ....   **(2 marks)**

Ebola

1   Which is the best way to control the spread of Ebola?

☐ **A** antibiotics

☐ **B** the hygienic disposal of faecal material

☐ **C** the spraying of water with insecticides

☐ **D** isolating victims for a week

> The disease is viral so check which options are relevant. For D, think about whether a week is going to be long enough.

(1 mark)

2   *Ebola is a deadly disease which:

- has an incubation period of 2–21 days
- involves diarrhoea
- damages body tissues, which can lead to bleeding from the eyes and other orifices
- is spread by droplets in air
- has a 50–90% death rate
- can be contracted from a dead body.

Ebola is caused by a virus and is extremely difficult to treat. It is being tackled by preventing its spread. Use the relevant facts above to explain strategies to prevent its spread.

> This is an extended writing question and needs to be planned logically. First decide which facts are relevant in preventing spread and then decide how to describe why they are important. For example, being spread by droplets in the air means that victims should be cared for by people who are wearing masks and their isolation needs to be such that air cannot escape to non-infected people.

....

....

....

....

....

....

....

....

....

(6 marks)

Ebola drug development

1   Explain what steps are taken to ensure a modern drug trial is valid.

....

....

....

....

....

.... **(3 marks)**

2   Explain why drugs are tested on animals and then on humans in a modern drug trial.

....

....

....

.... **(2 marks)**

3   (a)   Explain the difference between treatment and a vaccine for a disease such as Ebola.

....

....

....

.... **(2 marks)**

(b)   Assess which you think is the most useful to develop.

> In 'assess' questions, you must give careful consideration to all the factors or events that apply and identify which are the most important or relevant. Make a judgement on the importance of something and come to a conclusion where needed.

....

....

....

....

.... **(3 marks)**

The cell cycle

1   Describe the events that occur in mitosis from the start of prophase up to the end of anaphase.

> Think carefully about the detail here: plan by listing the stages, prophase, metaphase, anaphase and telophase, then describe each one in order. The logical sequence is very important here.

....

....

....

....

....

....

....

....   **(5 marks)**

2   The photograph shows cells in various stages of the cell cycle.

> The chromosomes are on the equator in A but at the poles in B but as yet no new nuclear membranes are being made.

(a)   Name the stages shown by A and B.

A is ....

B is ....   **(2 marks)**

(b)   Explain which part of the cell cycle is shown by cell C.

> This is an 'explain' question so it is not enough to just say what it is; you must say why it is.

....

....

....   **(2 marks)**

Roles of mitosis

Practical skills

1 The flowchart shows the stages in the preparation of a root tip squash to observe cells in the cell cycle.

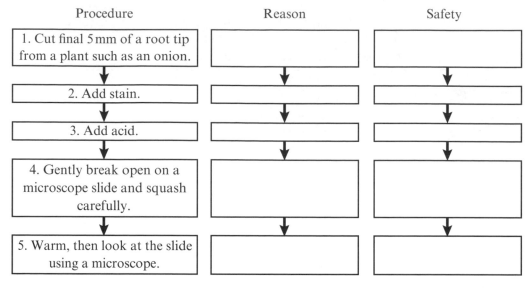

Procedure	Reason	Safety
1. Cut final 5 mm of a root tip from a plant such as an onion.		
2. Add stain.		
3. Add acid.		
4. Gently break open on a microscope slide and squash carefully.		
5. Warm, then look at the slide using a microscope.		

(a) Complete the flowchart with reasons for the procedures and safety comments.

> At stage 5 you will need two reasons as there are two procedures. For safety, you must be precise in your answer; vague statements like 'take care' will not gain marks.

(10 marks)

(b) Name a suitable stain at stage 2.

....   **(1 mark)**

Maths skills

2 After a zygote is formed during fertilisation, cell division occurs.

(a) Calculate the number of cells that will be present in the embryo after the first four divisions of the zygote.

> Remember that in every division each cell will become two new cells.

....   **(1 mark)**

(b) What kind of cell division is involved in the growth of the embryo?

....   **(1 mark)**

Guided

3 Greenfly reproduce asexually throughout the summer but sexually in the autumn ready for the winter. Give reasons why.

Greenfly reproduce asexually in the summer to reproduce faster so ....

....

....

....

....   **(2 marks)**

Meiosis

1   In meiosis which of the following
    is a correct function of meiosis I?

> Make sure you have understood the difference between chromosome and chromatid; it might be the case that neither number is maintained.

    ☐ **A**   the number of chromosomes is halved

    ☐ **B**   maintains the number of chromatids

    ☐ **C**   maintains the number of chromosomes

    ☐ **D**   generates variation

(1 mark)

2   There are two types of cell division: mitosis and meiosis. In the table below put a
    tick (✓) in the boxes to show which involve mitosis and / or meiosis.

> The question allows you to tick both mitosis and meiosis for any of the statements.

Statement	Meiosis	Mitosis
produces variation		
involves two divisions		
used in growth of tissues		
used in reproduction		

(4 marks)

3   The diagram shows a cell undergoing meiosis II. Complete the diagram by drawing
    the cell in anaphase II and explain what is occurring.

metaphase II

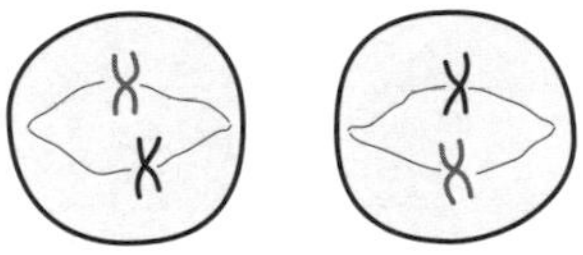

> In anaphase II there are still only two cells.

anaphase II

telophase II

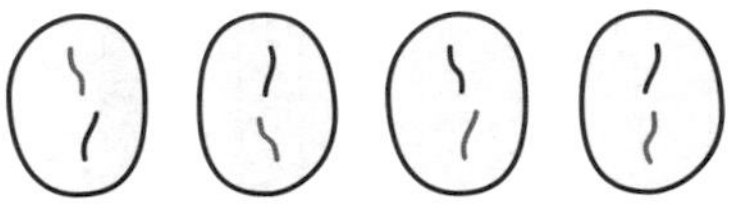

....

....

....

....

....

(4 marks)

Chromosome mutations

1 Down's syndrome is caused by a mutation. Which of the following best describes it?

☐ **A** monosomy ☐ **B** a translocation ☐ **C** polysomy ☐ **D** multiple allele **(1 mark)**

> **Guided**

2 (a) Describe how a trisomic zygote is produced.

During anaphase I of meiosis, one pair of homologous chromosomes may fail to

separate (non-disjunction) resulting in ....

....

....

....

.... **(5 marks)**

(b) What would differ in your description if the zygote developed into someone with Turner's syndrome?

> This question is asking for the difference so do not give the whole answer as in (a) but highlight the difference in producing monosomy rather than trisomy.

....

....

....

.... **(3 marks)**

3 The diagram shows two kinds of chromosome mutation.

(a) Explain what has happened in both cases.

....

....

....

....

....

....

....

.... **(4 marks)**

(b) Name the type of mutation. .... **(1 mark)**

Gametogenesis

1   The diagram shows a human sperm cell and a human egg.

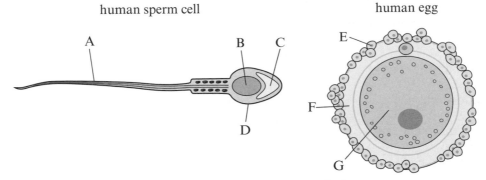

human sperm cell   human egg

Write the correct letter or letters next to the statements below.

A site containing an enzyme which digests the zona pellucida ....

A site with a haploid number of chromosomes ....

A site containing mitochondria ....

The zona pellucida .... **(4 marks)**

2   Complete the table below about mammalian sperm and eggs using ticks (✔).

> As well as their specialised features, sperm and eggs are eukaryotic cells.

Feature	Egg only	Sperm only	Sperm and egg	Neither sperm nor egg
mitochondria				
DNA				
cortical granules				
membrane				
cell wall				
diploid nucleus				
mid-piece				

(7 marks)

3   Write down the equivalent structure or structures found in oogenesis to those listed below for spermatogenesis. Include the numbers of each structure found.

spermatogenesis	oogenesis
one primary spermatocyte	
two secondary spermatocytes	
four spermatids	
four sperm	

(4 marks)

Plant sexual reproduction

1   In one species of flowering plant the chromosome number of each cell in the elongation region of the root is 20. State the number of chromosomes in the following:

> 20 is the diploid number 2*n*, so for each cell / nucleus you need to decide its '*n*' number.

A pollen mother cell   ....

An endosperm cell   ....

A pollen tube nucleus   ....

A generative nucleus   ....

A polar nucleus   ....   **(5 marks)**

2   State which of the following is correct in a flowering plant.

☐ **A**   The male gamete moves down the pollen tube.

☐ **B**   The generative nucleus divides by meiosis to form two male gametes.

☐ **C**   The generative nucleus divides by mitosis to form two male gametes.

☐ **D**   One male nucleus fuses with two female gamete nuclei to produce a triploid zygote.   **(1 mark)**

3   An investigation was undertaken to study the effect of a newly discovered hormone additive on the growth of pollen tubes. A large number of pollen grains was placed in a dilute sugar solution. Every six hours, for 30 hours, 500 pollen grains were removed and the length of the pollen tube of each was measured. The mean length of the pollen tubes was then calculated. This was repeated with the additive.

(a) Using the information in the graph, compare the mean pollen tube length in these two sugar solutions over this 30-hour period.

> This is a skills question and requires you to look at the patterns of the two lines and describe them.

....

....

....

....

....   **(3 marks)**

> Guided

(b) Explain the advantages to flowering plants of increased pollen tube growth.

If the pollen tube grows faster ....

....

....   **(2 marks)**

Effect of sucrose on pollen tube growth

1 Devise a procedure to determine the optimum sucrose concentration for the germination of the pollen grains shown in the photographs.

> You need to think about the control variables, what you will measure (the dependent variable), what you will vary, the range of what you will vary and safety / ethical issues.

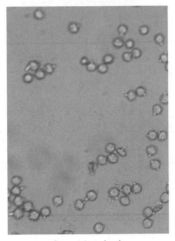

after 10 min in
1.2 moles per dm^{-3}
sucrose

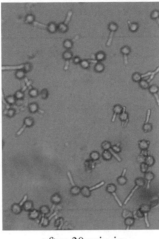

after 30 min in
1.2 moles per dm^{-3}
sucrose

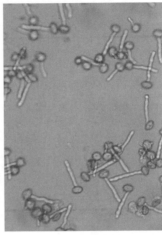

after 60 min in
1.2 moles per dm^{-3}
sucrose

....

....

....

....

....

....

.... **(5 marks)**

2 You are provided with a 2 M sucrose solution that also contains the mineral ions needed for pollen grain germination. Explain how you would make up a 0.5M solution of sucrose to use for encouraging your pollen grains to germinate.

....

....

....

....

....

.... **(2 marks)**

Had a go ☐ Nearly there ☐ Nailed it! ☐

Exam skills

1 (a) (i) The diagram shows a pollen grain growing on the stigma of a flower. Explain the role of enzymes in the growth of the pollen tube down the style.

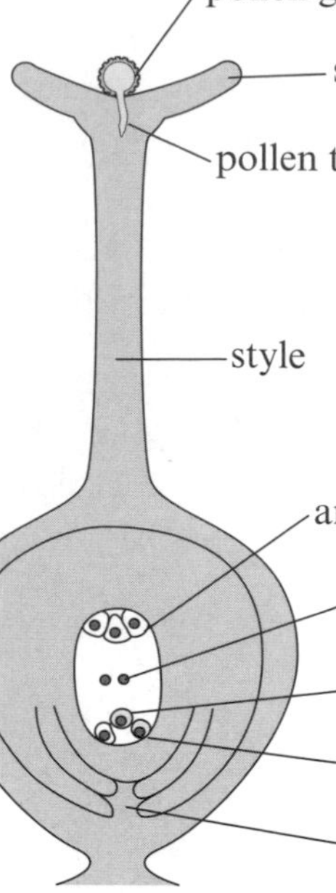
pollen grain
stigma
pollen tube
style
antipodal cell
polar nuclei
egg cell
synergid
micropyle

....

....

....

....

....

....

....

....

....

....

.... **(3 marks)**

(ii) Name the structures in the diagram which are haploid **and** fertilised by nuclei in the pollen grain.

> Certain words are often very important in questions. The emboldened 'and' is a case in point. In a real exam question it may not be emboldened so take care to read questions carefully.

....

....

....

.... **(2 marks)**

(b) The male gametes arise from meiosis and show genetic variation. One source of the variation is crossing over. The diagram on the left shows a pair of homologous chromosomes during meiosis. They are positioned next to each other but crossing over has not yet occurred. Complete the diagram on the right to show these chromosomes after crossing over has occurred.

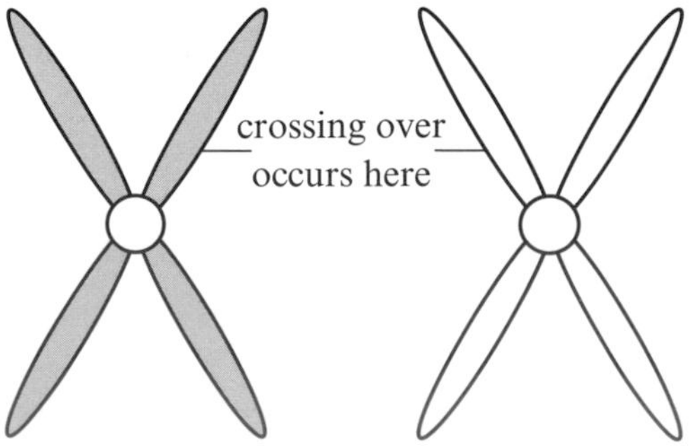
crossing over
occurs here

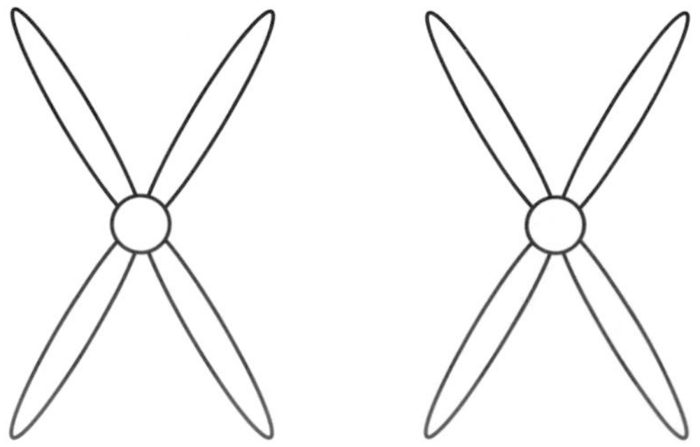

(1 mark)

The classification of living things

1   There is a hierarchy in the classification of living things. Which of the answers below is a correct part of that hierarchy?

> It is not only the order that is important. It is also important that there are no missing steps.

   ☐ **A**   Phylum, Kingdom, Class, Order

   ☐ **B**   Class, Order, Family, Genus

   ☐ **C**   Class, Order, Family, Species

   ☐ **D**   Order, Family, Species, Genus   **(1 mark)**

2   A new organism has been discovered in a research project exploring a tropical rainforest. Below is a list of some of its characteristics. Using this list, explain to which of the three domains, Archaea, Bacteria or Eukaryota, the organism belongs.

> There are **two** features that are unique to the domain of the organism.

Characteristic	Present/absent
mitochondria	absent
cell wall containing peptidoglycan	present
sensitive to antibiotics	present
may contain chlorophyll	present

....

....

....

.... **(2 marks)**

3   *At a recent scientific conference there was a debate as to whether two beetles which had identical morphology but occurred in two very different habitats were actually two different species. How would you investigate this issue by using reproduction?

> You will need to explain your proposed method and what would constitute the result in the case of either they are or they are not the same species.

....

....

....

....

....

....

....

.... **(6 marks)**

Molecular phylogeny

1 (a) A research paper stated that morphology, karyotype, host plant data and information from the analysis of patterns in DNA and RNA have been used to study the evolutionary relationships between species of fruit flies in the genus *Rhagoletis*. Which of the following techniques is likely to have been used to analyse patterns in DNA and RNA?

☐ **A** light microscopy

☐ **B** gel electrophoresis

☐ **C** PCR

☐ **D** electron microscopy

(1 mark)

(b) In the analysis of *Rhagoletis*, the gene for cytochrome oxidase II was analysed and found to have 687 nucleotides in all species studied. One of these species was a new one from Mexico. The researchers deduced that it was quite closely related to a North American species called *Rhagoletis pomonella*. They found that it had 22 nucleotides that differed from North American *Rhagoletis pomonella*. From this, they deduced how long ago it must have had a common ancestor with this species. To do this, they used data from another paper which suggested a nucleotide sequence divergence of 0.2% per 100 000 years. How long ago did they deduce the divergence took place?

> **Maths skills** Calculate the percentage change first by dividing the number of nucleotides that differ by the total number and multiply by 100. Then, knowing this and that the rate of change is 0.2% a year, the number of years can be found.

Answer .... years

(3 marks)

(c) In the study, DNA profiles were compared. Explain how this would have been done.

> This question is about how to compare the profiles **not** about how to obtain the DNA profiles.

....

....

....

....

....

(3 marks)

The validation of scientific ideas

1 Explain what is meant by the term peer review.

> Explain what a review involves and who a peer is.

....

....

....

....

....

.... **(3 marks)**

2 Until quite recently, most scientists accepted that all life was either prokaryotic or eukaryotic. However, in 1977, Carl Woese published a paper in a scientific journal in which he reported the use of molecular phylogenic evidence to suggest a third group, the Archaea. This is just one way of informing the scientific community of a discovery. State other ways in which he could have done this.

> State ways other than publishing a paper.

....

....

....

....

.... **(3 marks)**

3 Give two pieces of evidence, from our understanding of genetic material, which support Darwin's theory of evolution.

> This question is directing you to knowledge of DNA, which Darwin did not have.

....

....

....

.... **(2 marks)**

Evolution by natural selection

1   Wood ants are social insects that cooperate with each other in many ways, including finding and gathering food and building nests. The table below shows two possible adaptations of wood ants. Place a tick (✓) in the box or boxes that best describes whether the adaptation is behavioural, anatomical or physiological.

> Any particular adaptation can have more than one origin.

Adaptation	Behavioural	Physiological	Anatomical
production of formic acid as an alarm signal			
acting in a group to carry heavy prey to the nest			

(2 marks)

2   A disinfectant company claims that their product kills 99% of germs. Explain why this claim may not be reassuring to users of the product.

> This is an 'explain' question, so it is not enough to state your point – you must also give reasons or say why you have come to that conclusion.

....

....

....

....

....

.... (3 marks)

3   The development of antibiotic resistant bacteria is a major concern in the treatment of diseases. Describe the advice a GP might give to a patient with a bacterial infection to help reduce this general problem and explain why this advice is given.

> The advice given is to help reduce the development of antibiotic resistant bacteria, so it helps in the long-term and not the short-term treatment of the patient.

....

....

....

....

....

.... (3 marks)

Speciation and the evolutionary race

1   Over 75% of flowering plant species are thought to have arisen
due to autopolyploidy. This occurs when a mistake in cell division
causes a change in chromosome number. The old form and the
new form are immediately unable to breed with each other to
produce viable offspring. This is an example of:

> The word 'immediately' is very
> important in this question.

☐ **A** natural selection   ☐ **C** allopatric speciation

☐ **B** sympatric speciation   ☐ **D** artificial selection   **(1 mark)**

2   The hawthorn maggot fly (*Rhagoletis pomonella*) of the USA
feeds on hawthorn. However, since the cultivation of apple
orchards began in the USA in the mid-18th century, the fly
has been able to use apples as a food source. Laboratory
studies have shown that those flies completing their life cycle

> You will need to define sympatric
> speciation and then describe this
> particular fly's life cycle to show
> how it meets with the definition.

on apple (the *apple race* or breeding group) do so in under 40 days whereas those on
hawthorn (the *hawthorn race*) take over 50 days. In the wild, the apple race of flies
emerges about six weeks before the hawthorn race. The flies complete their entire life
cycle on the fruit wherever the female first laid her eggs. Analyse the information to
explain why this might be regarded as an example of sympatric speciation in progress.

....

....

....

....

....

....   **(5 marks)**

3   There are two species of rhinoceros in Africa. The white
rhinoceros and the black rhinoceros evolved from a
common ancestor. The white rhinoceros feeds on grasses. It
has a shoulder height of 1.5–1.8 m and has broad flat lips.
The black rhinoceros eats the leaves of shrubs. It has a
shoulder height of 1.4–1.7 m and has a pointed mouth.
Explain how these two species of rhinoceros evolved from their common ancestor.

> In questions like this where you
> are given a context, make sure you
> refer to the context provided and
> do not just write a generic answer.

....

....

....

....

....

....   **(5 marks)**

Biodiversity

1   In a study of birds in Borneo, species abundance was compared between two forest management strategies: sparing and sharing. The sparing strategy involves intensive logging of some plots and leaving others unlogged. The sharing strategy involves light logging of all plots. The data are shown in the graph. Analyse the data to comment on the relative effect of these two strategies on species abundance.

> The word 'analyse' means that you need to comment on the graph **and then** relate this to the situation being discussed to make a judgement on the issue.

....

....

....

....

....

....

....

(3 marks)

2   Pasture on limestone is a habitat with high biodiversity. The grazing of cattle and sheep suppresses the growth of scrub, allowing light penetration and maintaining a high diversity of small plants. Recently, reduced grazing has occurred in some places and this has allowed brambles and hawthorn to grow, leading to a reduction in biodiversity. Two possible management strategies have been proposed to conserve these habitats: by mowing or by controlled sheep grazing (conservation grazing). Proponents of conservation grazing believe this is the better method. Design an investigation to test this suggestion.

> Explain how you would carry out this investigation and how you would record and analyse the results. Compare the results using the index of diversity formula.

....

....

....

....

....

....

....

(5 marks)

Ex situ conservation: zoos and seed banks

1   State three concerns about keeping animals in zoos.

> Try to include some scientific concerns as well as animal welfare issues.

....

....

.... **(3 marks)**

2   *Explain the advantages of conserving plants by using seed banks.

> A star (*) next to a question means marks will be awarded for your ability to structure your answer logically, showing how the points that you make are related to or follow on from each other where appropriate.

....

....

....

....

....

....

....

....

.... **(6 marks)**

Guided

3   Some tiger species are threatened with extinction in the wild. Captive breeding programmes in zoos are trying to help with this problem. Two new tigers are introduced to the breeding stock every seven years to maintain 90% genetic diversity. Explain why it is necessary to maintain 90% genetic diversity to eventually allow reintroduction of the tigers into the wild.

90% genetic diversity keeps many alleles in the population, thus ....

variety of .... The population needs to show a lot of phenotypic variation as

the tiger lives in a wide range of .... in the wild. Also, if there were a change in

the environment, the tigers would be unlikely to be able .... if genetic variety

was .... **(4 marks)**

47

In situ conservation: habitat protection

1   Guatemala is a less economically developed country according to the IMF (International Monetary Fund) designation. The country suffered many wars and much unrest throughout the 20th century. The per capita income in Guatemala is $3807, whereas in the UK it is nearly $46 000 and in the USA nearly $56 000. Guatemala's Mayan Biosphere Reserve contains the largest remaining area of rainforest in Central America.

   (a)   Explain how the national parks and nature reserves can protect the rainforest.

   ....

   ....

   ....

   ....

   ....

   ....   **(3 marks)**

> Guided >

   (b)   Explain the likely purpose of the reserve's buffer zone.

   The buffer zone stretches across the whole of the southern edge of the reserve

   ....

   ....

   ....

   ....

   ....   **(3 marks)**

   (c)   Assess the likely success of this reserve.

   ┌──┐
   │ 'Assess' questions require you to identify important factors, make a judgement of │
   │ importance and reach a conclusion. You should consider both sides of the question and │
   │ come to a judgment. │
   └──┘

   ....

   ....

   ....

   ....

   ....

   ....

   ....   **(4 marks)**

Exam skills

**Maths
skills**

1 A study in Zimbabwe showed that areas with elephants have a higher biodiversity than those without. However, further work showed that a density of over 0.5 elephants km^{-2} leads to a reduction in biodiversity. The area studied was 66 000 km^2.

   (a) Calculate the elephant population in this area that would achieve maximum biodiversity.

   ....

   .... **(2 marks)**

   (b) *The chart shows elephant populations in this area over a 20-year period. An international law giving full protection to elephants was implemented in 1989. Analyse the information to explain the likely changes in biodiversity over this period.

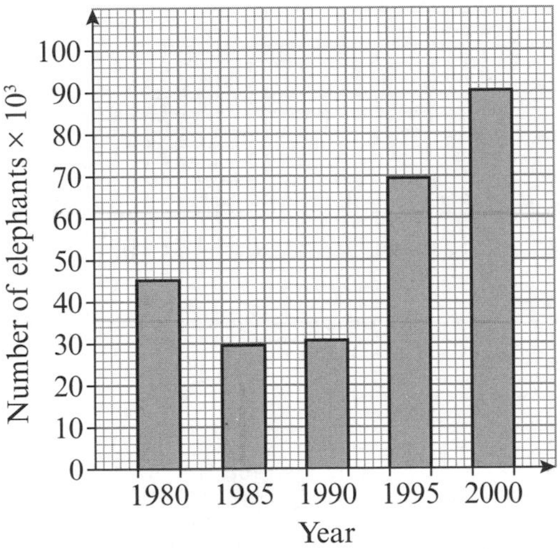

> The star (*) indicates that in your answer the points that you make should be related and follow on from each other. It is a good idea to plan your answer with this in mind. You are also asked to base your answer on your analysis of the data given. Make sure that you put some quantitative analysis into your answer to justify your statements.

   ....

   ....

   ....

   ....

   ....

   ....

   .... **(6 marks)**

   (c) Explain how an index of biodiversity could have been measured in each year of the study.

   ....

   ....

   ....

   ....

   ....

   .... **(4 marks)**

Had a go ☐ Nearly there ☐ Nailed it! ☐

The cell surface membrane

1 The cell membrane is made of a phospholipid bilayer. Explain why phospholipids form bilayers.

> This is an 'explain' question, so it is not enough to state your point – you must also give reasons or say why you have come to that conclusion.

> This is all about which part of the phospholipid molecule is hydrophobic and which is hydrophilic.

....

....

....

....

....

.... **(4 marks)**

Maths skills

2 The photograph shows the membrane between two cells. Calculate the thickness of the membrane between A and B. Express your answer in standard form as a fraction of a metre and in nanometres.

> Use the magnification formula:
> $\text{magnification} = \dfrac{\text{size of image}}{\text{size of real object}}$
> You will need to convert to metres and nanometres. Remember a mm is 10^{-3} m and a nanometre is 10^{-9} m.

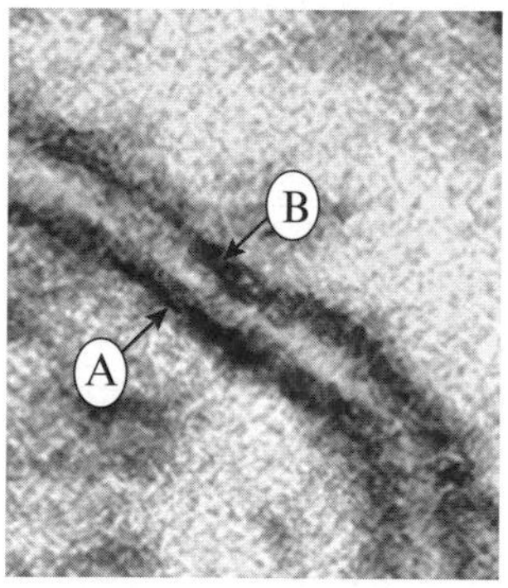

Magnification = 1×10^6

Answer .... **(3 marks)**

3 The diagram shows the fluid mosaic model of a cell membrane. Fully name structures A to D.

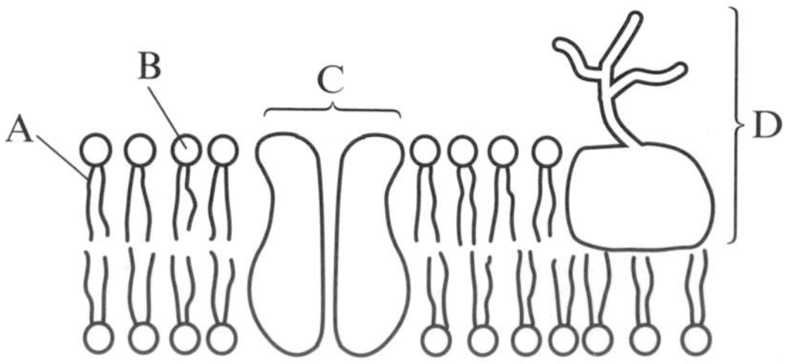

A is ....

B is ....

C is ....

D is .... **(4 marks)**

Membrane permeability

1 One piece of beetroot was placed in a tube containing 15 cm³ of water and left for
15 minutes. This was repeated for seven different concentrations of ethanol. A sample
of the fluid around the beetroot piece was placed in a colorimeter to determine the
intensity of red colouration of the fluid. The results are shown in the graph.

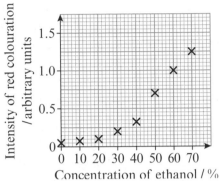

(a) State the colour of the filter that would have been used in the colorimeter.

.... **(1 mark)**

(b) State **two biotic** and **two abiotic** variables you would need to keep constant in
this investigation.

> Biotic will be to do with the living material; abiotic with the non-
> living environment in which the investigation is carried out.

....

....

....

....

....

.... **(4 marks)**

(c) Explain the result at 0% ethanol.

....

....

.... **(2 marks)**

(d) Explain the effect that increasing concentrations of alcohol have on beetroot
membranes as shown in the graph.

....

....

....

....

.... **(4 marks)**

Passive movement across membranes

1   An investigation was carried out into the permeability of a cell membrane to a number of different non-polar, organic molecules. The molecules differed in their size and in their solubility in oil compared to their solubility in water. The higher the solubility, the more soluble the molecule is in oil compared to water. The graph shows the results of this investigation. The size of the circle drawn on the graph indicates the size of the molecule: the larger the circle, the larger the molecule. A student concluded that the data support the fluid mosaic model of membrane structure. Analyse the data from this experiment to assess the validity of this conclusion.

> The word 'analyse' means that you need to comment on the data and then relate your comments to the situation being discussed to make a judgement.

> The data might support one aspect of the fluid mosaic model and not another, e.g. the hydrophobic phospholipid bilayer might be supported but what about protein channels?

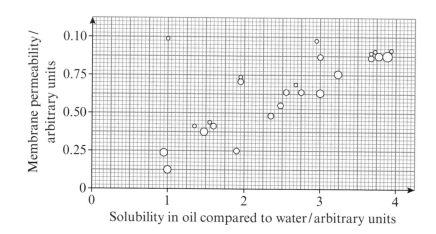

....

....

....

....

....

.... **(4 marks)**

2   The table shows some of the ways in which molecules can pass across membranes. Complete the table by writing **correct** or **incorrect** in each cell.

Process	Requires energy from respiration (ATP)	Requires a concentration gradient
passive diffusion		
facilitated diffusion		
osmosis		
active transport		

(4 marks)

Osmosis and water potential

1 The photograph shows some cells from the epidermis of a red onion which have been immersed in a solution.

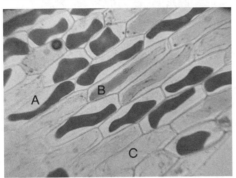

> You will need to deduce what the solution is that the cells have been in and then explain how you have come to this deduction.

(a) Explain the appearance of the cells labelled A, B and C.

....

....

....

....

....

.... **(4 marks)**

(b) (i) Deduce the turgor pressure that you would find in cell A.

> 'Deduce' means you need to work something out.

....

.... **(1 mark)**

(ii) Explain your answer to (i).

....

....

....

....

....

.... **(3 marks)**

(iii) State the name given to the appearance of cell A.

.... **(1 mark)**

Active transport

Guided 1 Compare and contrast transport across cell membranes by endocytosis and exocytosis.

> You must discuss both similarities and differences in 'compare and contrast' questions and look at both angles.

Both processes involve the use of .... to move contents

.... Both processes also require ....

Endocytosis moves .... but exocytosis transports

....

(3 marks)

2 An experiment is carried out to investigate movement across the membrane of a cell. The rate of movement of substances was followed over time. In one experiment, the membrane does not have specific channels for the substance. In another, there are carrier proteins – in one case with an inhibitor and in the other without. The graph shows the results.

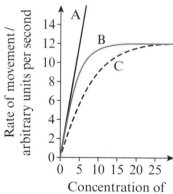

Which graph line shows the experiment with no specific channel?

Answer ....

Which graph line shows the results for movement via a channel protein with no competitor for the movement?

Answer ....

Which graph line shows the results for movement via a channel protein with a competitor?

Answer ....

Explain your answers.

> This is similar to enzymes and competitive inhibition – see p. 19.

....

....

....

....

(5 marks)

Surface area to volume ratio

Guided

1 An investigation was carried out into the relationship between size and rate of diffusion. Agar blocks (cubes of equal sides) of different sizes were placed in a dye and the time taken for the dye to reach the centre of the block was recorded. The table shows the block size and the times taken.

Length of side / cm	Area of surface of cube / cm^2	Volume of cube / cm^3	$\frac{SA}{V}$ ratio	Time for whole block to become coloured / seconds
13	1014			380
10		1000		300
7			0.86	100
5	150			53
3				20

(a) Complete the table by adding the area, volume and $\frac{SA}{V}$. **(5 marks)**

(b) Plot a graph of the relationship between $\frac{SA}{V}$ and time taken.

When plotting graphs, accuracy of positioning of the points is crucial. Make sure you also label axes with the name / variable and units.

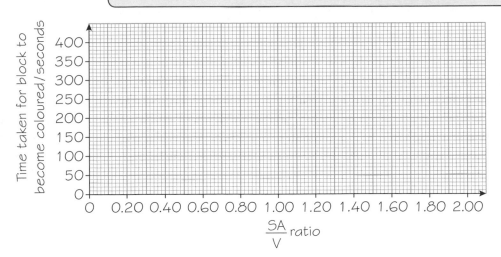

(4 marks)

(c) Analyse the data in your graph to support the hypothesis that small organisms are able to carry out sufficient gas exchange to meet their needs without specially adapted gas exchange surfaces such as gills or lungs.

Describe the trend in the data first then say how this affects the need for a gas exchange surface.

....

....

....

....

....

.... **(4 marks)**

Gas exchange in mammals

1 A lung condition commonly suffered by smokers is called emphysema. The diagrams show alveoli from the lung of a person with emphysema and alveoli from the lung of a healthy person. Use information from the diagrams and your own knowledge to explain the problems caused by emphysema.

alveoli from a lung of a person with emphysema

capillary

alveoli from a healthy person

> This question is all about reduction in surface area so describe the ways in which this has happened then describe the effect on gas exchange.

....

....

....

....

....

....

.... **(5 marks)**

Guided

Maths skills

2 A student wanted to find out by how many times the presence of air sacs, alveoli, in the lungs increases the surface area for gas exchange. He assumed that the lungs are spheres and have a radius of 89 mm giving a volume of 6 000 000 mm³ and also that the alveoli are spheres with a diameter of 0.25 cm. Calculate by how many times the presence of air sacs, alveoli, in the lungs increases the surface area for gas exchange.

volume of a sphere is $\frac{4}{3}\pi r^3$

so an alveolus = $\frac{4}{3} \times \pi \times 0.125^3 = 0.008$ mm³

> The number of alveoli in the lung is given by:
>
> $$\frac{\text{volume of lungs in mm}^3}{\text{volume of one alveolus in mm}^3}$$

therefore, number of alveoli in **two** lungs = ...

(4 marks)

Gas exchange in insects

1   In insects, blood flows out of the blood vessels and into the body cavity.
This is known as

☐ **A** a closed circulatory system

☐ **B** a single circulatory system

☐ **C** an open circulatory system

☐ **D** a double circulatory system   **(1 mark)**

2   Complete the table, which shows some of the adaptations of the gas exchange system of an insect.

Feature	Advantage
air sacs	
chitinous hoops in tracheae	
closable spiracles in body segments	
abdominal pumping movements	
non-chitinous tracheoles	

(5 marks)

3   (a)   Label the insect gas exchange system.

A is ....

B is ....

C is ....   **(2 marks)**

(b)   Describe how insects take oxygen into their bodies.

....

....

....

....

....   **(3 marks)**

Gas exchange in fish

1 The flow of water across the gills of fish is countercurrent, so water flows in the opposite direction to the flow of blood in the capillaries of the gill lamellae. The graphs show the changes in percentage saturation of the water and blood with oxygen as they pass over the lamellae for a countercurrent exchange system and a parallel exchange one. Analyse this information to explain the advantages of fish having a countercurrent exchange system, rather than a parallel exchange one.

> Think about the concentration of oxygen and when equilibrium will be reached.

Countercurrent
exchange system

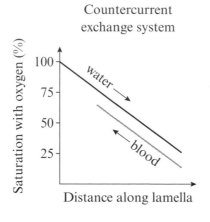

Parallel
exchange system

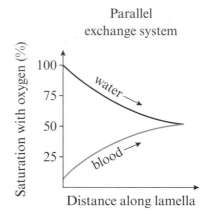

In the countercurrent exchange system, the blood flowing into the lamella is

meeting water that has not yet lost ....

....

....

.... **(3 marks)**

2 Explain why a fish dies quickly in air when air has much more oxygen than water (in which oxygen is relatively insoluble).

> Think about the structure of the gills and what might happen to the surface area of the gills in air.

....

....

.... **(2 marks)**

Gas exchange in plants

1   *The graph shows the results of a study of stomatal aperture, potassium ion (K^+) and sucrose concentration in guard cells. Glucose is made in the process of photosynthesis and then converted to sucrose for transport. Analyse the data and use your knowledge of stomata to assess the role of potassium ions and sucrose in stomatal opening and closing.

> Do **not** describe the curves in detail but look at the general trends, e.g. the rise in K^+ is followed by a rise in stomatal opening, which is then followed by a rise in sucrose.

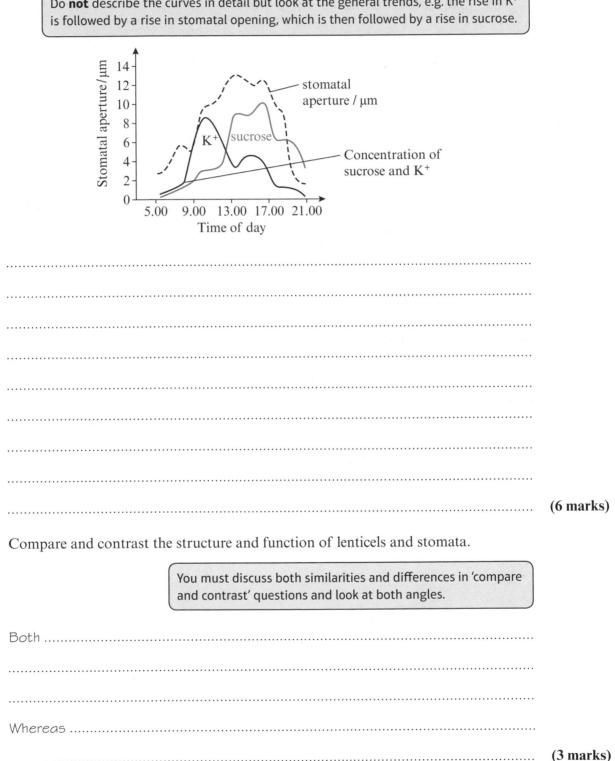

....

....

....

....

....

....

....

....   **(6 marks)**

Guided   2   Compare and contrast the structure and function of lenticels and stomata.

> You must discuss both similarities and differences in 'compare and contrast' questions and look at both angles.

Both ....

....

....

Whereas ....

....   **(3 marks)**

The heart and blood vessels

1 *Describe the structure of the mammalian heart.

> For 'describe' questions, you do not need to make a judgement or explain how or why, you only need to describe what is requested.

> There is a lot to say here; it might be best to make a list of each point and then put it together into prose to prevent rambling. You could also sketch a diagram.

....

....

....

....

....

....

....

....

.... **(6 marks)**

2 (a) State the function of valves in the circulatory system.

....

.... **(1 mark)**

(b) Explain why valves are found where they are in the circulatory system

> Think about the places where valves are and how blood is being moved along there – so why should blood move in the opposite direction, which is what the valves are there to prevent?

....

....

....

....

....

....

.... **(4 marks)**

Single and double circulation

1 The diagrams show the circulatory system of a mammal and the circulatory system of a fish. The arrows show the direction of blood flow. Using this information and your own knowledge, explain why the double circulatory system is needed by mammals but not by fish.

> Think about the difference between fish and mammals in two ways, i.e. fish are 'cold-blooded' and mammals are 'warm-blooded'; fish gills are in contact with water but mammal lung surface is in contact with air.

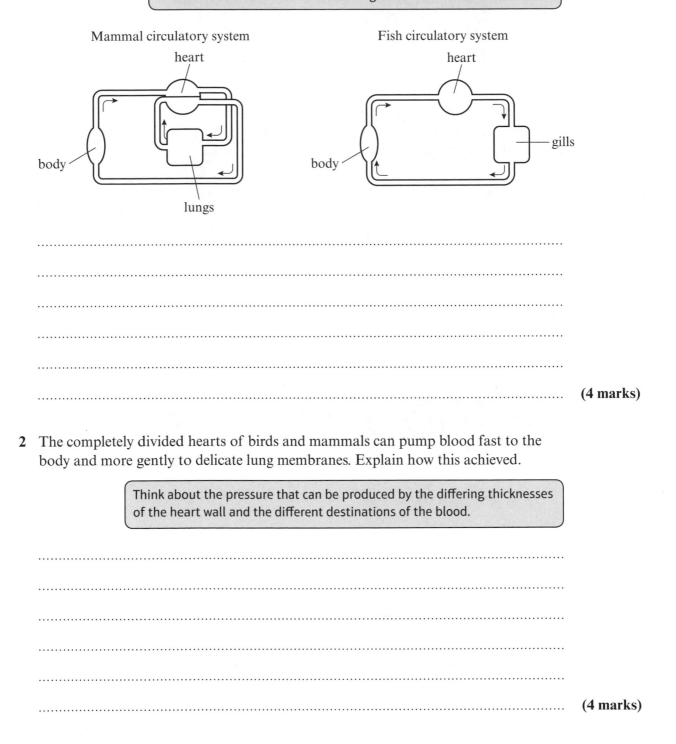

Mammal circulatory system

Fish circulatory system

....

....

....

....

....

.... **(4 marks)**

2 The completely divided hearts of birds and mammals can pump blood fast to the body and more gently to delicate lung membranes. Explain how this achieved.

> Think about the pressure that can be produced by the differing thicknesses of the heart wall and the different destinations of the blood.

....

....

....

....

....

.... **(4 marks)**

The cardiac cycle and the heartbeat

1   Describe how cells in the sinoatrial node (SAN) are involved in controlling heart rate.

....

....

....

....

.... **(3 marks)**

Maths skills

2   The graphs show pressure changes and an ECG during one cardiac cycle of the left side of the heart.

(a)   Calculate the heart rate.

> If you get a calculation question based on a graph you will almost certainly need to take readings from the graph, so make sure these are accurate. Check that you understand what both the x- and y-axis scales mean.

Heart rate ....   **(2 marks)**

(b)   After how many seconds does each of the following happen?

AV valve closes ....

Aortic valve closes ....

AV valve opens ....   **(3 marks)**

(c)   Describe what is happening during QRS and T phases of the ECG.

> In 'describe' questions, statements need to be developed as they are often linked but you do not need to include a justification or a reason.

....

....

.... **(2 marks)**

Blood and tissue fluid

1 Complete the table comparing blood plasma, tissue fluid and lymph.

	Blood plasma	Tissue fluid	Lymph
where found			in the lymph vessels
protein content			low to none
glucose content		variable	
cell content			

(4 marks)

Maths skills

2 Fluid enters or leaves blood capillaries depending on the balance of various forces. The diagram shows these forces in a capillary and its surrounding cells. The pressures are given in millimetres of mercury as used still in medicine, but SI units for pressure are pascals. 1 mm mercury = 133.3 pascals (Pa).

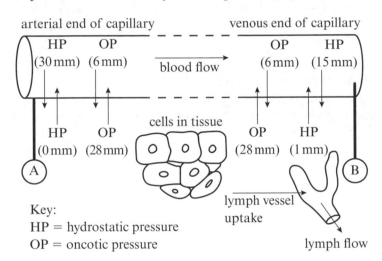

(a) Calculate the net pressures at A and B. Express your answers in Pa.

> A very important tool in maths calculations is the bracket ( ). Here, you need to work out the pressures 'out' **minus** those 'in', and there are two of each so you need to bracket them together.

(3 marks)

(b) Add arrows to the bars above A and B to show net direction of fluid movement.   **(1 mark)**

(c) Predict the consequences of a value at A of 10 mm and at B of −5 mm.

....

.... **(1 mark)**

Blood clotting and atherosclerosis

Guided 1 Describe the sequence of events in the clotting process.

A blood clot may form when a blood vessel wall ....

....

....

....

....

....

....

.... **(5 marks)**

2 Describe how atherosclerosis develops.

> This kind of topic could be set out as a flowchart of events, or bullet points – both are acceptable.

....

....

....

....

....

.... **(4 marks)**

3 Explain how atherosclerosis can cause the chest pain associated with an attack of angina.

....

....

....

....

.... **(4 marks)**

Haemoglobin and myoglobin

1   The graph shows the oxygen dissociation curve of haemoglobin (Hb) for three different concentrations of CO_2.

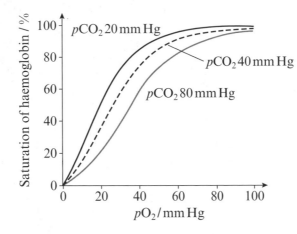

(a)   Calculate the percentage change in the carrying capacity of Hb for O_2 when CO_2 changes from a pCO_2 of 20 mm Hg to one of 80 mm Hg at a pO_2 of 40 mm Hg.

(3 marks)

(b)   State the name of this effect of CO_2 on Hb O_2 carrying capacity.

....   **(1 mark)**

(c)   Draw a line on the diagram to show the shape of a fetal Hb curve at a pCO_2 = 20 mm Hg.

> Think about problems like this logically; do not just try to remember what it is. In this case, the fetus needs to get oxygen from its mother so its Hb must have a higher affinity at any given pO_2.

(1 mark)

Had a go ☐ Nearly there ☐ Nailed it! ☐

Transport of water in plants

1 The diagram shows cells in part of a longitudinal section (LS) of a plant stem.
C and D represent below and above the section.

D
B
A
E
C
F

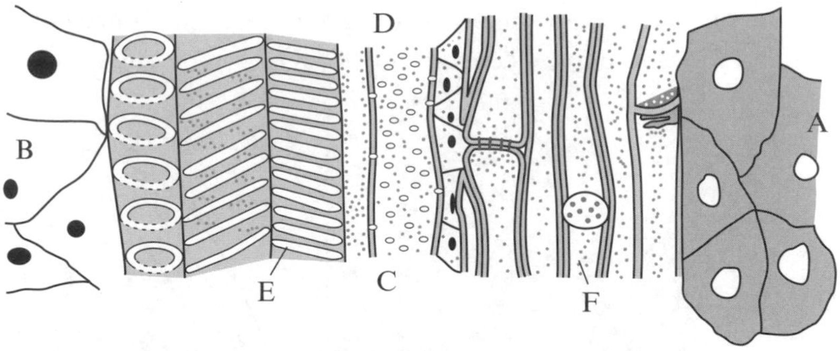

(a) (i) Which letter represents the place nearest to the exterior of the plant stem?

You will almost certainly have learnt the positions of these cells from a transverse section (TS) where you are looking at the stem end on. This is asking you to use that knowledge when looking side on (as above, which is an LS).

Answer .... **(1 mark)**

(ii) Which letter would be a place of very high sucrose concentration?

Answer .... **(1 mark)**

(b) Using the drawing and your own knowledge compare and contrast the structure and function of E and F.

'Compare and contrast' means to look for the similarities and differences of two (or more) things.

....

....

....

....

....

.... **(5 marks)**

2 Plants living in salty soil may take up sodium and chloride ions. Explain how they stop these ions from entering the xylem.

....

....

....

.... **(3 marks)**

3 Label the apoplastic and symplastic pathways on the diagram.

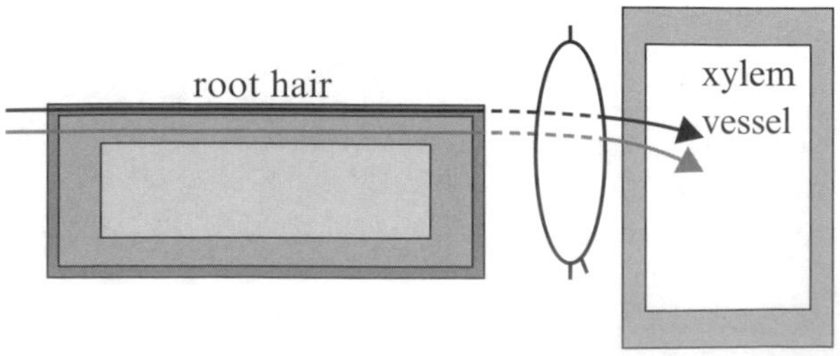

root hair
xylem
vessel

(2 marks)

The cohesion-tension model

1   Cavitation is quite common in tall trees. It is the breaking of water columns in the xylem of the tree. Cavitation can be detected by the sound (short click) made as the column snaps. The graphs show the number of cavitation events in a pine tree (EPM = event per minute), together with the light level and wind speed over two days. Analyse the information to explain what causes the columns to break.

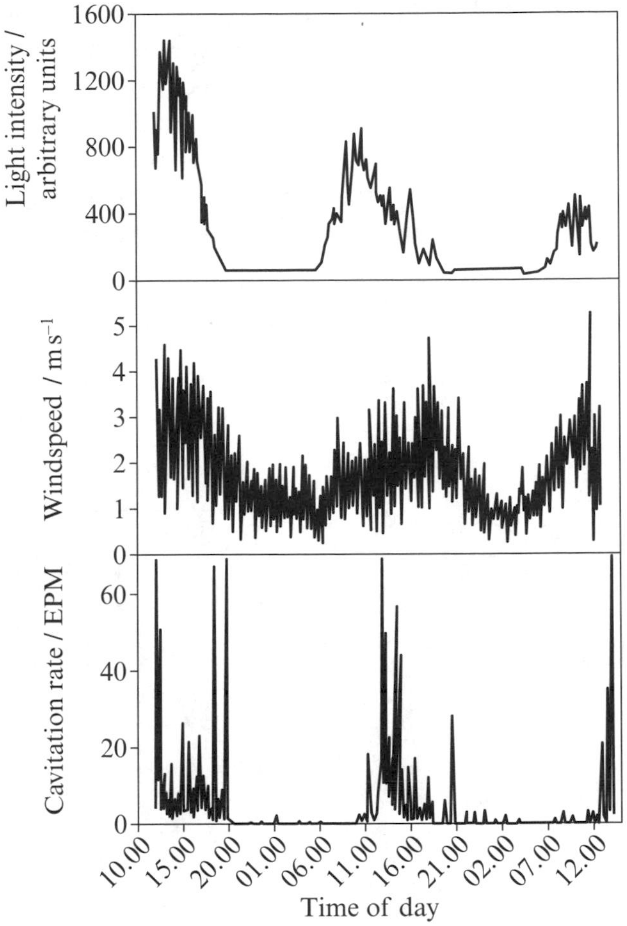

> You will need to put together what you can see in the graphs and what you know about factors affecting water loss and how this drives its movement in a plant.

....

....

....

....

....

....

....

....

....   **(4 marks)**

2   Compare and contrast the meaning of cohesion and adhesion in the context of plants and water.

....

....

....

....

....

....   **(4 marks)**

The mass-flow hypothesis

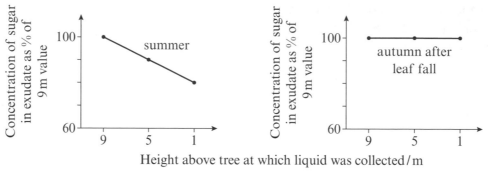

Guided 1 Cuts were made in the bark of an ash tree, causing the contents of the phloem (phloem exudate) to ooze out. The liquid was collected from different heights of the tree and analysed for sugar concentration. Two sets of observations were made: one in summer and another after leaf fall in the autumn. The results are shown in the graphs with sugar concentration given as a percentage of its concentration at 9 m from the ground. Analyse the data to explain how they support the theory that phloem transport is from a source to a sink.

The concentration of exudate is highest at the top of the tree in summer ....

....

....

....

.... **(4 marks)**

2 The following are facts about the transport of substances in the phloem.
   - Phloem cells are living and are supported by living companion cells.
   - Different solutes, for example sucrose and amino acids, move at different rates.
   - Substances may move in both directions in the same sieve tube.
   - The concentration of sucrose in the phloem near leaf cells is up to 30 times that in the leaf cells.
   - Phloem sap exudes from aphid mouthparts.

   Criticise the mass-flow hypothesis of transport in the phloem using this information and your own knowledge.

> 'Criticise' means inspect a set of data, an experimental plan or a scientific statement and consider the elements. Look at the **merits** and **faults** of the information presented and support judgements made by giving evidence.

....

....

....

....

....

....

....

.... **(5 marks)**

The uptake and loss of water

1 Explain why it is possible to get an accurate estimate of transpiration rate by measuring the uptake of water by a leafy shoot.

> Transpiration is the loss of water from a leaf by evaporation.

....

....

.... **(2 marks)**

Practical skills

2 The graph shows the results of an experiment to determine the effect of light intensity on the transpiration rate from the upper and lower surfaces of a leaf.

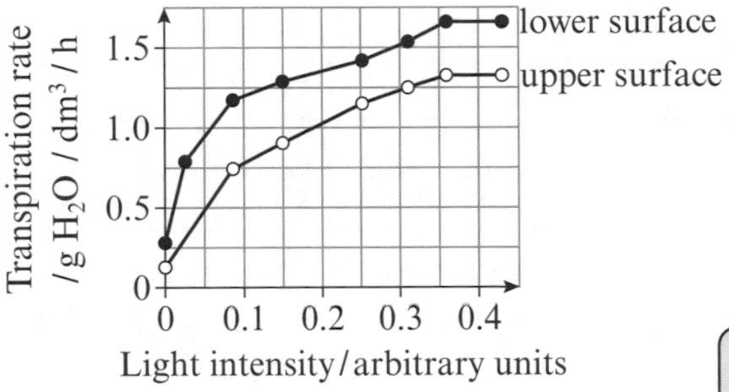

> This relates to Core practical 8.

(a) Name three controlled variables in this experiment.

....

....

.... **(3 marks)**

(b) Choose one of the variables you have named in (a) and explain how the results would be affected if it was not controlled.

....

....

.... **(2 marks)**

(c) Analyse the data in the graph to compare and contrast the effects of light intensity on the rate of transpiration from the two surfaces of the leaf.

> How are they the same; how are they different? 'Analyse' means that you need to comment on the data and then relate your comments to the situation being discussed to make a judgement. Also, when asked to analyse some data, it is often a good idea to do some manipulation on it. Here, rates are higher from the lower surface, but how much higher might be worth working out.

....

....

....

....

.... **(3 marks)**

69

Exam skills

1 (a) Xylem vessels transport water from the roots to the leaves with the help of cohesion. Explain what is meant by cohesion in this sentence.

....

....

.... **(2 marks)**

 Maths skills

(b) An experiment was carried out to find the value of Ψ for some plant tissue. The following results were obtained.

Sucrose concentration / mol dm⁻³	Percentage change in mass of potato chip
0.0	20
0.2	10
0.4	−3
0.6	−17
0.8	−25
1.0	−30

(i) Plot a graph of these data and use it to determine Ψ for this tissue.

> When plotting graphs, accuracy of positioning of the points is crucial. Make sure you also label axes with the variable and units.

(5 marks)

(ii) A thin section of the tissue in the 0.6 mol dm⁻³ solution was cut and viewed through a microscope. Draw a labelled diagram of the likely appearance of a single cell from this section.

> When drawing diagrams, make sure your lines are clear and not feathered. Labelling lines should be drawn with a ruler and end on the object they label. In this question, you need to think about the state of the cell in this solution.

(3 marks)

Exam skills

1 The diagram shows a section through the heart of a mammal.

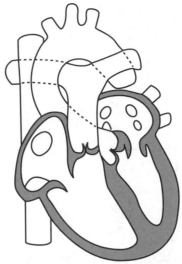

(a) Deduce from the diagram which stage of the cardiac cycle this heart is in.

> **Guided**

Both sets of atrioventricular valves are ....

....

....

.... **(3 marks)**

(b) ECGs can be used to diagnose abnormalities in the heartbeat. One such
abnormality is a ventricular ectopic beat. This occurs when a region of the
ventricle has a similar effect on the heart as the sinoatrial node (SAN).

The diagrams below show a normal ECG trace and a trace that shows a ventricular ectopic
beat, labelled E. The traces were recorded from left to right. Changes in blood pressure in
the pulmonary artery are shown over the same period of time.

Normal heart
ECG

Heart with ectopic beat
ECG

E

Change in blood pressure
in pulmonary artery / kPa

Change in blood pressure
in pulmonary artery / kPa

Time / seconds

Time / seconds

*Use this information to explain the effect of the ectopic beat on heart activity.

....

....

....

.... **(3 marks)**

Stages of aerobic respiration

1 (a) Complete the table by inserting ticks (✓) in the appropriate boxes to show in which stage of aerobic respiration these processes occur.

	Glycolysis	Link reaction	Krebs cycle	Oxidative phosphorylation
ATP produced				
ATP hydrolysed				
CO_2 produced				
NAD reduced				
Reduced FAD is oxidised				
Occurs in cytoplasm				
Occurs in mitochondria				

(7 marks)

(b) The diagram shows a stage of cellular respiration which occurs in the cytoplasm of an animal cell. Describe the process by which hexose bisphosphate is produced from glucose.

> For 'describe' questions, you do not need to make a judgement or explain how or why, you only need to describe what is requested.

> Think about where the phosphate may have come from and how it could be transferred to glucose.

glucose → hexose bisphosphate → 3-carbon compound

....

....

....

....

.... **(3 marks)**

(c) Explain how molecules produced during the conversion of hexose bisphosphate to a 3 carbon compound are important in the electron transport chain. **(3 marks)**

....

....

....

....

Glycolysis

1   The diagram shows two molecules involved in one of the stages in glycolysis.

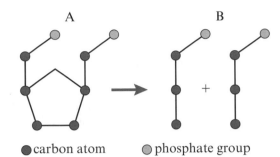

● carbon atom   ○ phosphate group

(a)   Name molecules A and B.

A is ....

B is ....   **(2 marks)**

(b)   Name the processes described below.

> Remember OIL RIG: Oxidation Is Loss, Reduction Is Gain.

ATP production in glycolysis

....

Addition of phosphate

....

Removal of hydrogen atoms

....

Addition of hydrogen atoms to NAD

....   **(4 marks)**

(c)   Explain why glucose needs to be phosphorylated and what happens to the phosphate groups in later reactions of glycolysis.

> This is an 'explain' question, so it is not enough to state your point – you must also say why you have come to that conclusion.

....

....

....

....

....

....   **(3 marks)**

Link reaction and Krebs cycle

1 The diagram shows some stages in a metabolic process in cells.

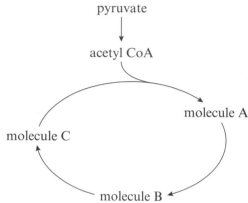

pyruvate

↓

acetyl CoA

molecule A

molecule C

molecule B

(a) State the location(s) of these processes.

.... **(1 mark)**

(b) There are other molecules produced by these metabolic processes.
Name these other products.

> The number of marks is not always a guide to how many points
> you need to give. In this case there are four points you could make.

....

....

....

.... **(2 marks)**

> Guided

(c) Explain the changes in carbon atom number between molecules A, B and C.

> Do not forget the contribution
> of acetyl CoA.

> Try to make two further points
> to complete this answer.

Acetyl CoA is a 2-carbon molecule. It binds to the 4-carbon molecule C to form

molecule A which is a 6-carbon molecule. ....

....

....

....

.... **(4 marks)**

Oxidative phosphorylation

1   This diagram shows some structures found in a mitochondrial inner membrane.

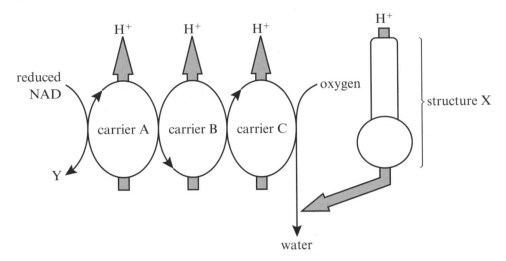

(a) Name compound Y.

....   **(1 mark)**

(b) Which part of aerobic respiration
     does the diagram show?

> Note that the question asks for
> 'part' and not 'stage'.

☐ **A** electron transport chain

☐ **B** glycolysis

☐ **C** Krebs cycle

☐ **D** link reaction   **(1 mark)**

⟩ **Guided** ⟩   (c) Explain the roles of carriers A, B and C and the structure labelled X.

> Name structure X in your answer.   Aim to make four more points.

Electrons pass from carrier A to B to C in a series of redox reactions. This

releases energy which is used to ....

....

....

....

....

....

....   **(5 marks)**

Anaerobic respiration

1   The diagram shows some of the stages of glycolysis.

Stages of glycolysis

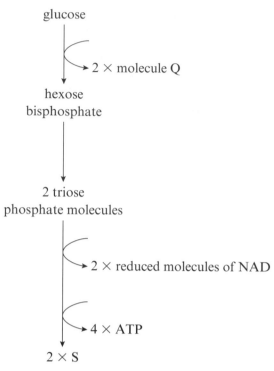

glucose

2 × molecule Q

hexose
bisphosphate

2 triose
phosphate molecules

2 × reduced molecules of NAD

4 × ATP

2 × S

(a) What is molecule Q?

☐ **A** ADP   ☐ **B** ATP   ☐ **C** oxidised NAD   ☐ **D** reduced NAD   **(1 mark)**

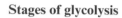 **Guided**

(b) In anaerobic conditions in an animal cell, molecule S will not enter the link reaction. Explain what happens to molecule S and the importance of these processes to an animal cell.

> Oxidative phosphorylation will not occur in the absence of oxygen but the cell still needs ATP. How will it produce some ATP?

Molecule S is pyruvate and it is reduced by the addition of hydrogen atoms to

form lactate. These hydrogen atoms come from reduced NAD................................

....

....

....

....   **(3 marks)**

(c) Compare and contrast anaerobic respiration in animal and plant cells.

> You must discuss both similarities and differences in 'compare and contrast' questions. For example, what molecule is reduced?

....

....

....

....

....   **(3 marks)**

Using a respirometer

1

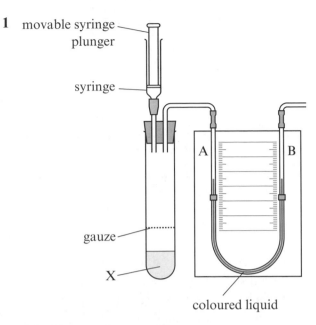

movable syringe
plunger

syringe

A   B

gauze

X

coloured liquid

(a) Name substance X.

.... **(1 mark)**

> **Guided**

(b) *Describe how this apparatus could be used to find the mean rate of aerobic respiration of peas.

> Lots of enzymes are involved in respiration. It is important to control the temperature so it does not affect the rate of respiratory enzyme action. Make sure you structure your answer logically, showing how the points you make are related. You should also support your points with biological facts.

The apparatus should be placed in a water bath at 25 °C and left to equilibriate.

....

....

....

....

....

....

....

.... **(6 marks)**

(c) This apparatus was used to investigate the rate of respiration in woodlice.
The coloured dye moved 5.2 cm in 3 minutes. The diameter of the capillary tube was 3 mm. Calculate the rate of oxygen uptake.

> Change cm into mm. Remember you need to multiply the area of the end of the capillary tube by how far the dye has moved!

Rate = ... **(2 marks)**

Photosynthetic pigments

1 Chloroplasts contain several pigments that absorb light energy of differing wavelengths.

(a) State where these pigments would be found in a chloroplast.

.... **(1 mark)**

(b) Explain the benefit to the plant of having several different pigments.

> Aim to give three points here.

....

....

.... **(3 marks)**

(c) An experiment was carried out where algal cells were placed in tubes with an indicator that changed colour at different pH ranges. The colour of the indicator at different pH ranges is shown below. Each tube was wrapped in a filter which allowed limited wavelengths of light into the tubes and left for 3 hours. The results of the investigation are shown below.

yellow		orange		red		magenta		purple
pH 7.6	pH 7.8	pH 8.0	pH 8.2	pH 8.4	pH 8.6	pH 8.8	pH 9.0	pH 9.2

Type of filter on bottle	Colour of light allowed into the bottle	pH value
clear acetate	all (400–680 nm)	9.0
primary red	red (600–680 nm)	8.6
bray blue	blue (425–500 nm)	8.4
primary green	green (475–590 nm)	7.8
black paper	none	7.6

(i) Explain why one tube was covered in a black filter.

....

.... **(2 marks)**

(ii) Explain the change in pH for both the green and black filters.

> Think about why chloroplasts look green. How much light energy is being absorbed?

....

....

....

.... **(3 marks)**

Investigating photosynthesis

1 An investigation was carried out into how light intensity affects the rate of photosynthesis in algae. Algal cells were captured in alginate balls and placed in an indicator which changes colour as shown in the diagram. The experiment was set up as shown and left for two hours. The colour of the indicator in the tubes was then measured using a colorimeter to measure light absorbance.

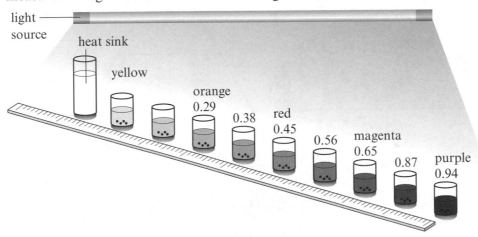

The light source also radiated heat energy which could have affected the rate of photosynthesis in the nearest tubes. A heat sink, consisting of a glass container filled with distilled water, was used in this experiment to absorb heat energy from the light source and control the variable of temperature.

(a) State two variables that should have been controlled in this experiment.

number of algal cells in each ball

.... **(1 mark)**

(b) Relative light intensity is calculated using the formula $\frac{1}{D^2}$

Use this formula to calculate the relative light intensity for 250 cm.

Distance from lamp (D) / cm	Relative light intensity (×10⁻⁵)	Absorbance / arbitary units
250		0.93
350	0.81	0.82
500	0.40	0.65
780	0.16	0.54
1250	0.06	0.44

.... **(1 mark)**

(c) Explain the trends that can be seen in these data.

> Aim to make at least two more points.

Algal balls in the tubes closer to the light source received light at higher intensity. More light energy was absorbed by the photosynthetic pigments in the chloroplasts.

....

....

....

.... **(2 marks)**

79

Chloroplasts

1   The image shows an electron micrograph image of a chloroplast.
    The magnification is ×5430.

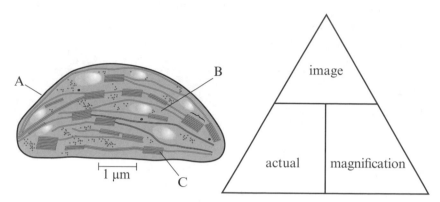

 Maths skills   (a) Calculate the actual size of this chloroplast.

        Answer .... μm   **(2 marks)**

  (b) (i) State the letter for the site of the photosystems.

         Answer ....

> Photosystems absorb light energy.

        (1 mark)

      (ii) State the letter that shows the part of the
         chloroplast where enzymes for
         the light-independent reaction would be found. ....   **(1 mark)**

  (c) Chloroplasts contain polysaccharide grains, DNA and ribosomes.

      (i) Name the polysaccharide that can be found inside chloroplasts.

        ....
            (1 mark)

      (ii) Describe how this polysaccharide is formed.

        ....

        ....

        ....
      (2 marks)

▷ Guided ▷   (iii) Explain why the chloroplast contains DNA and ribosomes.

        Both the light-dependent and light-independent reactions involve reactions

        catalysed by enzymes. ....

        ....

        ....

        ....

        ....
      (2 marks)

Photosynthesis: light-dependent stage

1 (a) State the location of the light-dependent stage of photosynthesis in the chloroplast.

.... **(1 mark)**

(b) Complete the table below by inserting ticks (✓) in the appropriate boxes to show which processes occur in cyclic and which occur in non-cyclic photophosphorylation.

	Cyclic photophosphorylation	Non-cyclic photophosphorylation
Electrons are passed along an electron transport chain		
Redox reactions involved		
Electrons raised to higher energy level		
Electron from a water molecule replaces electron lost from chlorophyll		
The same electron replaces electron lost from chlorophyll		
NADP reduced		
ATP produced		

(7 marks)

> **Guided**

(c) *The graph shows the results for an investigation carried out into the effect of increasing herbicide concentration on the growth of mung beans. The herbicide inhibited electron flow between photosystems. Explain why this would result in a decreased rate of plant growth.

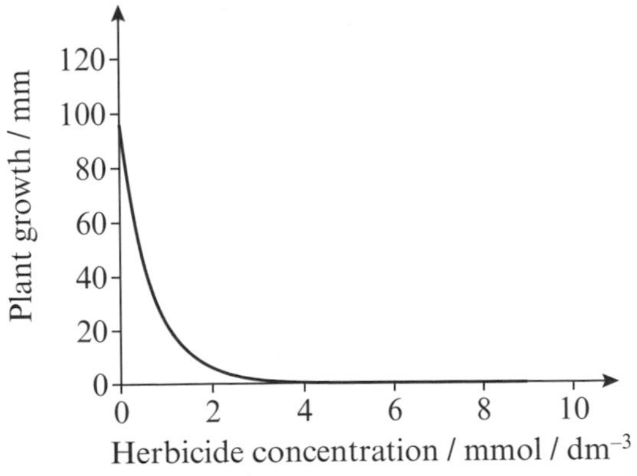

> This question is asking you to relate the effect of the lack of the light-dependent stage to the effect on the light-independent stage of photosynthesis.

The herbicide would reduce the flow of electrons along electron carriers in the electron transport chain. This would result in less ATP and reduced NADP being formed. ....

....

....

.... **(6 marks)**

Photosynthesis: light-independent reactions

1 (a) The light-independent stage of photosynthesis uses some of the products of the light-dependent stage.

   (i) Name the two products of the light-dependent stage that are used in the light-independent stage of photosynthesis.

....

.... **(2 marks)**

   (ii) State the location in the chloroplast of the light-independent stage.

.... **(1 mark)**

> **Guided**

   (b) Describe how carbon dioxide molecules can be reduced into molecules of GALP.

   > Aim to make two more points.

Carbon fixation occurs when the RuBisCO enzyme catalyses a reaction between

carbon dioxide and RuBP. Molecules of GP are formed. ....

....

....

....

.... **(4 marks)**

   (c) The glucose produced in photosynthesis can be used to form other molecules such as amino acids. What elements might be added to glucose in order to form an amino acid?

   > This is a synoptic question. You need to recall information from Topic 1 about the structure of biological molecules.

....

....

.... **(2 marks)**

   (d) Starch can be formed from the glucose molecules produced in the light independent stage of photosynthesis. Describe the structure of starch.

....

....

....

....

.... **(4 marks)**

Limiting factors in photosynthesis

> **Guided**

1 Bedding plants such as pansies are often grown by garden centres in greenhouses. This allows them to maximise productivity by increasing the rate of photosynthesis and plant growth. Describe three factors that could be controlled in the greenhouse and explain how each could increase the rate of photosynthesis.

> Temperature has been done for you so give two more factors with an appropriate explanation.

An increase in temperature up to the thermal optimum would increase the rate of the photosynthetic enzymes and therefore increase the rate of photosynthesis.

....

....

....

....

.... **(4 marks)**

2 High oxygen concentrations can reduce plant yield. The effect of oxygen concentration on the rate of carbon dioxide absorption was investigated.
Explain why the yield from plants is decreased at higher concentrations of oxygen.

> Oxygen is complementary to the RuBisCO active site.

....

....

....

....

....

....

.... **(5 marks)**

Exam skills

1   The diagram shows some processes that occur inside a leaf cell.

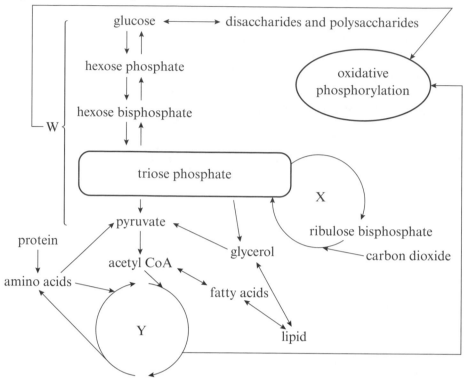

(a)   Complete the table to give the name of reaction pathways W, X and Y and
place ticks (✓) in the appropriate boxes to show the location of the process in
the leaf cell and the biological process in which it occurs.

	Name of reaction pathway	Location in leaf cell	Occurs in photosynthesis	Occurs in respiration
W				
X				
Y				

(3 marks)

(b)   The compounds from some of these processes are used in oxidative
phosphorylation. State the three products of oxidative phosphorylation.

....

.... **(2 marks)**

(c)   CoA is a coenzyme involved in carrying a 2-carbon molecule called
acetate to the Krebs cycle. Explain the function of three more
coenzymes that are involved in the processes shown in the diagram.

> What reduces all
> three coenzymes?

....

....

....

....

.... **(4 marks)**

Culturing microorganisms

1 Draw a line from each type of medium to its correct description.

> Make sure you only draw one line between each of 1, 2 and 3 to just one of A, B or C.
> Any ambiguous answers will be marked as incorrect.

Medium

| 1 nutrient broth |
| 2 selective medium |
| 3 agar plate |

Description

| **A** solid medium containing nutrients needed for microorganisms to grow |
| **B** solution of nutrients in which microorganisms grow |
| **C** can be used to grow specific types of microorganisms as it contains substances that promote growth of desired microorganisms and substances to inhibit the growth of other microorganisms |

(1 mark)

2 Which of the following is **not** needed for bacterial growth?

☐ **A** nutrients

☐ **B** suitable temperature

☐ **C** suitable pH

☐ **D** UV light

(1 mark)

3 Which of the following statements are true?

i   In schools and colleges, inoculated Petri dishes should be cultured at 30 °C or below.

ii Aseptic technique ensures all cultures and inoculated plates are sterile.

iii When inoculating a nutrient agar plate, always use a flamed loop or a pre-packed sterilised plastic loop.

☐ **A** statements i, ii and iii

☐ **B** statements i and ii only

☐ **C** statements i and iii only

☐ **D** statements ii and iii only

(1 mark)

> **Guided**

4 Describe how you could find out, experimentally, the optimum temperature for the bacterium *E. coli* to grow on agar plates.

> You need to grow the bacteria at a range of stated suitable temperatures. Describe what medium you will use, how you will inoculate (refer to aseptic technique) and what you would expect to see.

> For 'describe' questions, you do not need to make a judgement or explain how or why, you only need to describe what is requested.

Use aseptic technique to inoculate nutrient agar plates or nutrient broth tubes.

....

....

....

....

....

(3 marks)

Growth of bacteria

 Maths skills

1 The diagram shows the central counting area of a haemocytometer grid and one of the 25 squares with some bacterial cells, seen under the microscope with a total magnification of ×1000. Each of the 25 squares is subdivided into 16 smaller squares. The solution containing the bacteria was a 0.0001 (10^{-4}) dilution.

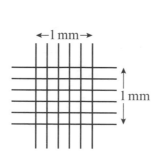

←1 mm→

1 mm

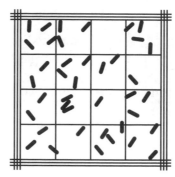

(a) How many bacteria are present in the one of the 25 squares shown here?

> Remember to consider ways of standardising your count, for example the rule about not counting those on the bottom or right hand side boundaries. Always show your working because if you have the right idea but make a silly slip you can still get some marks. A wrong answer with no working gets no marks.

.... **(1 mark)**

(b) The volume above one of the 25 squares (16 smaller squares) is 0.004 mm³ (= 0.004 µl)

(i) Calculate the number of bacteria in 1 µl of this diluted sample.

....

.... **(2 marks)**

(ii) Calculate the number of bacteria in 1 cm³ of this diluted sample.

> There are 10^3 µl in 1 cm³.

.... **(2 marks)**

(iii) Calculate the number of bacteria in 1 cm³ of the original undiluted sample.

> The answer to (iii) should be a much bigger number than your answer to (ii).

....

.... **(2 marks)**

 Maths skills

(c) Calculate the actual length of one of these bacteria.

> You have been told the magnification, so remember actual size = magnified size divided by magnification factor. Measure in mm and convert to µm. There are 1000 µm in 1 mm. Questions about magnification can come into any question on any suitable topic as it is an underlying core principle.

....

.... **(2 marks)**

Streak plating and measuring population growth

1 In an investigation, *E. coli* bacteria were grown at 40 °C in nutrient broth at pH 7. At 30-minute intervals a sample was taken and placed in a cuvette. This was then placed in a colorimeter (a cheaper version of a spectrophotometer) and green light shone through it. The absorbance (optical density) was read. The higher the reading, the more cloudy (turbid) the broth. This turbidity is directly proportional to the number of bacterial cells in the broth. After taking the colorimeter reading, 1 cm³ of the liquid culture was added to 9 cm³ distilled water and shaken. Using aseptic technique, a loopful of the diluted culture was spread onto a nutrient agar plate and incubated at 30 °C for 24 hours. The number of colonies on the agar plate was then counted. The table shows the data obtained.

Time / minutes	Number of colonies
60	50
90	
120	205
205	over 200; too many to count

The graph plots the data obtained from the turbidity (optical density) readings.

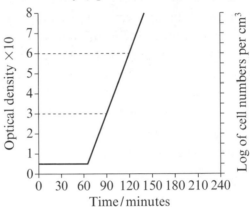

Guided

(a) Why is a logarithmic scale used for cell numbers on the second *y*-axis?

Because the range is too .... **(1 mark)**

(b) Fill in an estimate for the missing number in the table of data.

> The time is half way between 60 and 120 minutes so think about how the bacterial numbers might change during that interval.

.... **(1 mark)**

(c) At 60 minutes what was the number of bacterial cells per cm³ in the undiluted sample?

.... **(2 marks)**

(d) The generation time (*g*) is the time taken for a population of bacteria to double in size, during the exponential growth (log) phase. Use the graph to find the generation time for *E. coli* grown under the conditions of this investigation.

> There is one mark for a correct number and another mark for the correct units. Always include units.

Generation time is .... **(2 marks)**

(e) Explain how you would expect the data to differ from those described above if the nutrient broth tubes were incubated at 20 °C.

....

.... **(3 marks)**

Bacteria may cause diseases

1 To which kingdom do bacteria belong?
☐ **A** Fungi
☐ **B** Protoctista
☐ **C** Prokaryotae
☐ **D** Plantae

(1 mark)

2 Decide whether the following statements are true (T) or false (F) and circle the appropriate letter.

Exotoxins are lipopolysaccharides.	(T)	(F)
Exotoxins pass out of bacterial cells.	(T)	(F)
Endotoxins are made by Gram negative bacteria.	(T)	(F)
Exotoxins can damage red blood cells.	(T)	(F)

(4 marks)

> **Guided**

3 Sepsis (toxic shock) occurs when the body's non-specific immune system goes into 'overdrive' and mounts a huge inflammatory response. If not diagnosed quickly and treated immediately with antibiotics, it usually leads to multiple organ failure and death. Symptoms are headache, fever, feeling very cold, and sometimes a rash. Sepsis happens because the infecting bacteria produce exotoxins, and without antibiotics the specific immune system fails to combat the infection. If antibiotics are given quickly, often intravenously, then patients recover. However, in many cases the sepsis is not diagnosed.

(a) Explain how swift antibiotic treatment may combat sepsis.

Antibiotics kill the infecting bacteria so they cannot produce the exotoxins. Without these, the action of the non-specific immune response is stopped or reduced greatly and therefore ....

....

....

.... **(3 marks)**

(b) Discuss why it is difficult for doctors to diagnose sepsis.

> This is a question where you are asked to use your knowledge and the information in the stem of the question. Think about what doctors use to diagnose a condition – the symptoms the patients has. Look at the symptoms of sepsis given here – are they similar to symptoms of any other diseases?

> For 'discuss' questions, you should consider all aspects of the information, identify issues and explore them and give reasoned explanations.

....

....

.... **(2 marks)**

4 State two ways, other than by ingestion, by which bacteria can enter the body.

....

.... **(2 marks)**

Action of antibiotics

1 Many people have a sub-clinical infection of the bacterium *Mycobacterium tuberculosis*. They have some of the bacteria inside their lungs but they have no symptoms for many years. Eventually they may develop symptoms of tuberculosis (TB). Explain why people who are human immunodeficiency virus (HIV) positive may eventually show symptoms of TB.

....

....

.... **(2 marks)**

2 Compare the properties and actions of **bacteriostatic** and **bactericidal** antibiotics.

> If you are asked to compare two things, state any similarities as well as differences between them.

....

....

....

....

....

.... **(3 marks)**

Guided ▷ **3** Penicillin is an antibiotic that has been widely used for about 70 years.

(a) How does penicillin act on bacterial cells?

> For answering (a) and (b) you need to recall your knowledge of how each antibiotic works. Give as much relevant correct biological knowledge as you can and use correct technical terms.

Penicillin acts on bacterial cells by preventing growing Gram positive bacteria

from synthesising their cell walls. ....

....

....

....

.... **(4 marks)**

(b) How does tetracycline act on bacterial cells?

....

....

.... **(2 marks)**

Antibiotic resistance

1   The mutation rate in bacteria is about the same as it is in many other organisms but we can see the effects of those mutations more quickly because bacteria have such a short generation time. In addition to this, because bacteria do not have linear paired chromosomes, any mutation in a particular gene can be seen in the phenotype. As a result of mutation, some bacteria can produce beta-lactamase enzymes that catalyse the breaking open of a beta-lactam ring in the molecular structure of penicillin. Penicillinase is a specific type of beta-lactamase. It catalyses the hydrolysis of the beta-lactam ring.

The gene for penicillinase is present on plasmids of some bacteria. Penicillinase was first identified in 1940 in *E. coli* bacteria but during the next few years many other bacteria developed the ability to produce penicillinase. *E. coli* bacteria are part of the normal gut microbiota (bacteria that live in the gut).

> Guided

(a)   Explain the following terms:

> These definitions may not be on the Revision Guide page but you should understand them from other areas of your biology studies and be able to apply them in any context.

Mutation: ....

Phenotype: visible characteristic(s) of an organism

Hydrolysis: ....

Enzyme: ....

Plasmid: ....   **(4 marks)**

(b)   Explain why the lack of linear paired chromosomes in prokaryote cells leads to more frequent expression of gene mutations compared to expression of gene mutations in eukaryote cells.

....

....   **(2 marks)**

(c)   Explain why penicillinase cannot confer resistance to the antibiotic tetracycline.

....

....   **(1 mark)**

> Guided

(d)   Suggest how using antibiotics incorrectly leads to the spread of antibiotic resistance in populations of different species of bacteria.

> This requires you to apply all of your relevant biological knowledge and principles.

If someone takes antibiotics to kill a pathogen, the antibiotics could also kill some of their gut bacteria and select resistant bacteria in their gut.
These bacteria pass out in faeces and when they come into contact with other bacteria can pass their resistance genes by ....

....

....

....

....   **(4 marks)**

Other pathogenic agents

1   Some species of bacteria are pathogenic (they cause disease). Some viruses, fungi and protists are also pathogenic. The events outlined below relate to how an influenza virus may infect epithelial cells of the respiratory tract of the host (infected organism).

   A   Host cell lyses (splits) and dies.

   B   An antigen on the surface of the virus coat fits into a receptor on the host cell surface membrane.

   C   Droplets containing virus particles are breathed in.

   D   Host cell is 'tricked' into allowing the virus to enter.

   E   Many new virus particles are released.

   F   The viral nucleic acid directs the infected host cell to make new virus particles.

Arrange these events in the correct sequence.

> In order to gain both marks, all events must be in the correct order.

....   **(2 marks)**

2   Explain why specific viruses can only infect certain cells.

....

....

....   **(2 marks)**

3   State **two** ways, other than by droplets, that the influenza virus can enter a new host.

> If you are asked for two ways, do not give more. If you do then only the first two will be marked so make these the most important ways. If you gave three – one correct, one incorrect and one correct – you would only get one mark. You are not allowed to give a list and let the marker choose the correct answers from it!

....

....   **(2 marks)**

> **Guided**

4   Explain why fungal pathogens of plants are economically important.

> Things are economically important if they make money or if they cost money. Here you are asked to explain why. Aim to add another point.

They are important because they can infect crop plants, damaging their cells and tissues and absorbing nutrients from the plant. This weakens the plant and reduces its growth, therefore reducing its yield for human consumption. This leads to loss of profit for growers and less food availability to feed the human population. Agrichemicals may have to be used to reduce the spread of the virus and

....

....

....   **(5 marks)**

Malaria

1   Which of the following statements are true?

i   Female *Anopheles* mosquitoes are vectors of malaria.

ii   The infecting agent that causes malaria is a protozoan parasite called *Plasmodium*.

iii Malaria can be passed directly by contact from infected human to another human.

☐ **A**   statements i, ii and iii

☐ **B**   statements i and iii only

☐ **C**   statements i and ii only

☐ **D**   statements ii and iii only

(1 mark)

2   Female mosquitoes may take up *Plasmodium* cells when they bite an infected human. *Plasmodium* organisms then multiply in her stomach and migrate to her salivary glands. When she bites another human, several days later, she injects saliva with an anticoagulant into the wound and *Plasmodium* parasites may also enter the new wound. They enter the human liver and stay there for 7–10 days, producing no symptoms but multiplying within the liver cells. They then break out of the liver cells, go into the bloodstream and enter erythrocytes. They reproduce further and break out of the red blood cells, destroying these cells, every 2–3 days, producing intermittent fever.

(a) Why do only female *Anopheles* mosquitoes transmit malaria?

....

.... (2 marks)

There is no effective vaccine yet against *Plasmodium* but people can take prophylactic medicines, to prevent infection. The drugs enter the person's bloodstream and can kill any *Plasmodium* that enter via a mosquito bite. Travellers to an area where malaria is endemic take the tablets before travelling, while they are in the area and up to two weeks after returning home.

(b) Suggest why travellers must continue taking their anti-malarial tablets after returning home even if they have no symptoms of malaria.

| Read the information above to deduce the answer. |

....

....

.... (2 marks)

(c) Suggest why malaria patients suffer from anaemia.

....

....

....

....

....

| 'Suggest' means you need to apply your knowledge and understanding to a situation that you may not have covered. |

(2 marks)

Controlling endemic diseases

1   What does endemic mean in the context of disease?

   ☐ **A**   a sudden outbreak of an infectious disease

   ☐ **B**   an outbreak of an infectious disease that spreads worldwide across continents

   ☐ **C**   a disease that is always present in an area or population

   ☐ **D**   the study of patterns of disease transmission   **(1 mark)**

> **Guided**

2   Explain **two** methods of reducing the risk of catching malaria while in an area where malaria is endemic.

> You are asked to **explain** two methods so for each you need to describe the method and then say how it works. Do not give two similar answers. The first one here is about avoiding being bitten. The second answer should be about controlling mosquito numbers.

Avoid coming into contact with infected mosquitoes by not spending time near stagnant water, such as lakes and ponds, especially at dusk and night when mosquitoes bite and then you will not be bitten and will not receive the *Plasmodium* parasite.

....

....   **(2 marks)**

3   *Discuss the problems associated with using pesticide sprays to kill mosquitoes.

> Think about the evolution of resistance in mosquitoes and explain the underlying biological principles; think about how pesticides may affect other organisms in the ecosystem. Never use the word immunity when you mean resistance – the underlying mechanisms of the two processes are very different.

> Make sure you structure your answer logically, showing how the points you make are related or follow on from each other where appropriate. You should also support your points with relevant biological facts and / or evidence. Write in full sentences, not bullet points and use technical terms where appropriate.

....

....

....

....

....

....

....

....   **(6 marks)**

The role of white blood cells

1   Which group of cells is involved in the non-specific (innate) immune system?

☐ **A** macrophages and neutrophils

☐ **B** macrophages, natural killer cells and
        B lymphocytes

☐ **C** macrophages, neutrophils and natural killer cells

☐ **D** neutrophils, natural killer cells and T lymphocytes   **(1 mark)**

> Macrophages and neutrophils are phagocytic. Natural killer cells kill infected cells.

2   Which cells secrete many antibodies into the bloodstream?

☐ **A** T lymphocytes

☐ **B** natural killer cells

☐ **C** B lymphocytes

☐ **D** plasma cells   **(1 mark)**

3   All cells in the human body, except red blood cells, have MHC antigens on their
    surface. Red blood cells have other types of antigens, such as A or B antigens.
    Explain the following terms:

(a) antigen

....

.... **(2 marks)**

(b) MHC

....

.... **(2 marks)**

⊳ Guided ⊳   4   The table shows some of the leucocytes of the immune system. Complete the table
              by inserting ticks (✓) to show the functions and characteristics of these leucocytes.

Cells	Stimulate other cells involved in specific immune response	Stop the immune response when pathogens are destroyed	Produce antibodies	Contain large amount of rough endoplasmic reticulum	Remain in the body conferring long-term immunity
B cells					
T_h cells	✓				
T_s cells					
T_k cells					
plasma cells					
memory cells					✓

(6 marks)

The humoral immune response

1   Complete the following passage about the humoral specific immune response by filling in the blanks. Use words from the list below. Each may be used once, more than once, or not at all.

> endocytosis   antigen presenting cell   macrophages   mitosisa   T helper   cytokines
>
> memory   phagocytosis   antibodies   T suppressor   immune   B cells   plasma cells

When some bacteria enter the human body, they are ingested by the process of

.... by .... Inside this type

of cell, the bacterium is digested and its surface antigens are passed along the

endoplasmic reticulum into vesicles that are moved to the cell surface membrane.

The bacterial antigens are then displayed on the surface of the macrophage coupled

to MHC class 2 and alogside MHC class 1. The macrophage now acts as an

.... This cell presents antigen to ....

cells and when one with suitable receptors on its surface comes into contact with the

APC, it is stimulated to divide by .... This results in a

clone of T_h cells and T memory cells. The T_h cells secrete chemical messengers

called .... that stimulate a selected B cell to divide by

.... APCs also present antigen to B cells and those with

the complementarily shaped antibodies on their cell surface are selected. The antigen

may be enveloped by the B cell membrane by .... and the

B cell also acts as an APC. The selected B cell divides many times by

.... and produces a clone of B effector cells and a clone of

B .... cells. The effector cells differentiate into plasma cells

that are larger than B lymphocytes as they have large amounts of endoplasmic

reticulum and ribosomes. Plasma cells synthesise and secrete large amounts of

Y-shaped immunoglobulin proteins called .... These

.... bind to antigens on the surface of other pathogenic

bacteria of the same species, making it easier for .... to

ingest and destroy them. When all the invading pathogenic bacteria have been

destroyed, ... cells stop this immune response

.... cells remain in the body for a long time so if the same

pathogen invades again, a very swift immune response is mounted and the pathogen

is destroyed before any symptoms arise.

The person is .... to that pathogen.   **(8 marks)**

The cell-mediated immune response

1   State whether each of the following statements is true (T) or false (F) by circling the correct answer.

T_k cells bind to and destroy pathogens.   (T)   (F)

Infected host cells present the pathogen's antigens on their cell surface membrane.   (T)   (F)

The cell-mediated response involves APCs.   (T)   (F)

T_k cells release chemicals that cause infected cells to lyse.   (T)   (F)

T_k cells can destroy cancer cells.   (T)   (F)

T_k cells have receptors that fit the displayed pathogen's antigens and receptors that fit the host cells' MHC antigens.   (T)   (F)

The cell-mediated response does not involve clonal selection.   (T)   (F)   **(7 marks)**

2   The human immunodeficiency virus, HIV, contains RNA and because it also contains an enzyme, reverse transcriptase, which directs the infected host cell to make DNA using the viral RNA as a template, it is a retrovirus. It belongs to a subgroup of retroviruses called lentiviruses because it can remain dormant in the host for many years. HIV can infect many types of cell including T helper cells that have CD4 receptors on their cell surface membranes. Scientists think that this virus spread to humans from non-human primates early in the 20th century. Diseases that spread from animals to humans are called zoonoses. The influenza virus is an orthomyxovirus. It contains RNA but it is **not** a retrovirus.

(a)   Explain why HIV only infects cells with a certain type of receptor on their cell surface membrane.

....

....

....   **(2 marks)**

(b)   Suggest why the influenza virus is not classed as a retrovirus.

....

....   **(1 mark)**

(c)   Suggest why the viral disease Ebola is also classed as a zoonosis.

....

....   **(1 mark)**

> Guided >   3   Complete the table below by inserting ticks (✓) in the appropriate places.

Immune response	B cells involved	T_h cells involved	T_k cells involved	Antigen presentation	MHC involved	Clonal selection	Clonal expansion	Mitosis
cell mediated								
humoral	✓							

(2 marks)

Types of immunity

> **Guided**

1 Complete the table to describe the types of active and passive immunity.

	Active immunity	**Passive immunity**
Natural	person suffers from a disease caused by an infecting agent and mounts an immune response that gives long-term immunity after recovery	
Artificial		antibodies against a specific antigen found on the surface of a particular pathogen are injected into a person to give quick acting but short-lived immunity

(6 marks)

2 What is meant by 'herd immunity'?

....

.... **(2 marks)**

Maths skills

3 The measles vaccine is 95% effective. This means that only 95% of those vaccinated become immune. To achieve herd immunity, 95% of the population must be immune to measles in order to break the train of transmission from one human to another within a community or population. What proportion of the population should be vaccinated against measles to achieve 95% immunity?

Answer ....% **(1 mark)**

4 In 1960, the World Health Organization (WHO) set out to eradicate smallpox worldwide. They achieved this by 1980. They could not vaccinate everyone in many parts of the world but sent teams to areas where there were cases of smallpox and used ring vaccination – all contacts of the affected individual were vaccinated. The smallpox virus was stable (did not mutate) and the vaccine could be freeze dried and did not need to be refrigerated. The vaccine was given using a bifurcated needle placed just under the skin so many people could be trained to deliver the vaccine. Use the information above to suggest how the stability of the smallpox virus helped achieve the eradication of smallpox.

....

....

....

.... **(3 marks)**

Exam skills

Guided 1 Complete the table below by adding ticks (✓) where appropriate to show if a statement is true for an organism. (Two have been completed for you. Use a tick (✓) as directed and not a cross (✗).)

Microorganism	Not made of cells	Contain nucleic acid	Cells have a nucleus	May produce antibiotics	May be killed or inhibited by antibiotics
viruses					
fungi					✓
bacteria				✓	

(3 marks)

2 The graph shows changes in the number of cells and changes in the amount of DNA present in a milk culture of the bacterium *L. delbrueckii*, subspecies *bulgaricus*, over a period of 10 hours in anaerobic conditions at 30 °C.

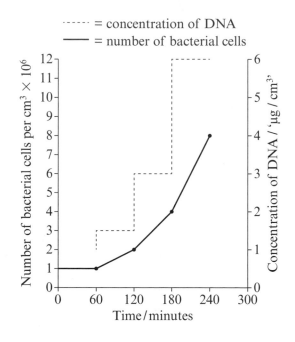

- - - - = concentration of DNA
——— = number of bacterial cells

(a) What is the generation time (doubling time) for this population of *L. delbreuckii* during the log phase?

.... (2 marks)

(b) Describe and explain the relationship between the two curves shown on the graph.

....

....

....

.... (3 marks)

(c) This bacterium *L. delbreuckii* is used to produce yoghurt. *L.d. bulgaricus* changes milk proteins to amino acids which are then used by another species of bacterium, *S. thermophilus*. Both species produce lactic acid. How would the pH of the milk culture change as the numbers of bacteria increased?

> Do not write 'it would become acidic' as you have been asked about the pH which has a numerical value.

.... (1 mark)

(d) By which process do bacterial cells divide?

☐ **A** mitosis   ☐ **B** meiosis

☐ **C** binary fission   ☐ **D** nuclear fission   (1 mark)

Using gene sequencing

1   What is a genome?

☐ **A** all the chromosomes of an organism

☐ **B** all the genes within a population

☐ **C** all the genetic material present in a cell or organism

☐ **D** all the DNA in the nucleus of a eukaryotic cell   **(1 mark)**

Guided 2   Complete the table below by adding a tick (✓) if DNA is present or a cross (✗) if DNA is absent from the organelles.

> Think about whether the cells have these organelles.

Type of cell	Organelle				
	Nucleus	**Ribosomes**	**Plasmids**	**Chloroplasts**	**Mitochondria**
prokaryotic	✗				
eukaryotic	✓				

(2 marks)

3   Arrange the following stages involved in the process of gene sequencing into the correct sequence.

A   DNA fragments are separated into single strands of DNA (ssDNA).

B   DNA molecules are amplified using the polymerase chain reaction (PCR).

C   ssDNA is replicated by adding nucleotide bases and terminator bases tagged with fluorescence.

D   DNA molecules are cut into fragments using restriction endonuclease enzymes.

Correct sequence: ....   **(1 mark)**

4   During the PCR, double-stranded DNA is first heated to separate it into single strands.

(a) Which type of bonds in the DNA molecules are broken during this stage?

....   **(1 mark)**

(b) Give a reason why the polymerase enzyme used in the PCR is obtained from the bacterium *Thermous aquaticus* that lives in hot springs.

> 'Give a reason' means you may need to think about it and apply your biological knowledge.

....

....

....   **(2 marks)**

Transcription factors

1 What happens when cells differentiate?

☐ **A** Some genes are destroyed.

☐ **B** Some chromosomes are coated with RNA and deactivated.

☐ **C** Some genes are activated and expressed.

☐ **D** All genes are transcribed.

(1 mark)

2 The diagram shows the action of transcription factors.

B ....   C ....

D ....

A ....

E ....

F ....

protein

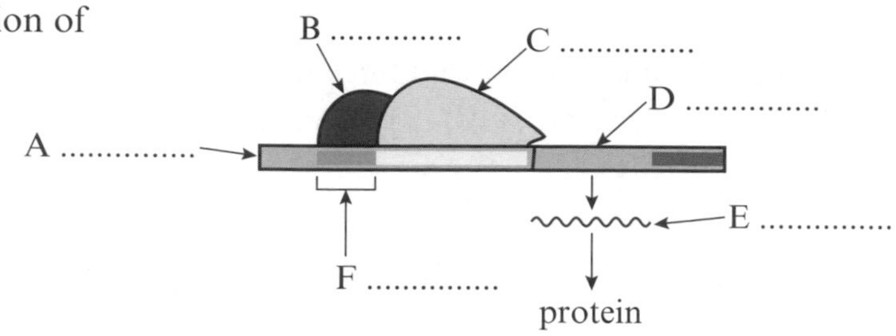

(a) Label the diagram by choosing from the list below.

- mRNA
- gene
- enhancer region
- promoter region
- RNA polymerase

- transcription factor
- DNA polymerase
- double-stranded DNA
- single-stranded DNA
- tRNA

(6 marks)

(b) On the diagram indicate the structures made of protein and the structures made of nucleic acids.

> When asked to label a diagram, make sure you include clear and accurate labels and use a ruler where necessary.

(2 marks)

3 What is a transcription factor?

☐ **A** a protein that binds to DNA to initiate DNA replication

☐ **B** a protein, not part of RNA polymerase, that initiates the formation of mRNA

☐ **C** an enzyme that catalyses the formation of mRNA

☐ **D** a region of DNA upstream of the gene that activates the promoter region

(1 mark)

4 Explain how transcription factors bind only to specific base sequences within a length of DNA.

> This is an 'explain' question, so it is not enough to state your point – you must also give reasons or say why you have come to that conclusion.

....

....

....

(2 marks)

Guided 5 State four examples of transcription factor control.

1 ....

2 ....

3 ....

4 Pathogens change gene expression in host cells so the pathogens can infect the host cell.

(3 marks)

RNA splicing and epigenetics

1   Match each of the following with its correct description.

    A   a section of a gene that is expressed and translated into protein

    B   a molecular machine made of RNA and protein that removes introns

    C   a section of a gene that is transcribed but not translated into protein

    D   transcribed mRNA containing both introns and exons

    Intron is ....   Exon is ....   Spliceosome is ....   Pre-mRNA is ....   **(4 marks)**

2   Which of the following is **not** an example of epigenetic control?

    ☐ **A** methylation of cytosine molecules within a gene

    ☐ **B** deacetylation of histone proteins in chromatin

    ☐ **C** a substitution of the nucleotide base cytosine by guanine

    ☐ **D** a long ncRNA molecule that smothers and inactivates one of each of the pairs of X chromosomes in cells of females   **(1 mark)**

> **Guided**

3   How might histone deacetylase inhibitors be used to treat people with epilepsy and memory problems?

Acetylation of histone proteins in chromatin reduces gene expression. ....

....

....

....

....

.... **(4 marks)**

Maths skills

4   A gene is 3000 base pairs long and the protein produced, when it is expressed, contains 650 amino acids, including the initial one, methionine.

    (a) How many base pairs are in all the introns of this gene?

    Answer .... base pairs   **(3 marks)**

    (b) How might this gene be able to be expressed as several other different proteins besides this one?

    .... **(1 mark)**

Stem cells

1   Which type of stem cell can produce any type of cell?

   ☐ **A** induced pluripotent

   ☐ **B** multipotent

   ☐ **C** pluripotent

   ☐ **D** totipotent

   (1 mark)

Guided   2   For each of the following statements, indicate whether it is true or false.

   Stem cells can continue to divide by mitosis.   True

   Pluripotent stem cells can produce a new embryo. ....

   Adult stem cells occur in bone marrow. ....

   Totipotent and pluripotent stem cells are only present in embryos. ....

   Plant cells do not have any epigenetic regulation of protein production. ....

   Differentiated human cells contain fewer genes than a human zygote does. ....

   Epigenetic changes to gene expression may or may not be inherited. ....

   Epigenetic changes to gene expression involve alterations to the base pair

   sequence. ....

   Differentiated cells cannot express all of their genes. ....

   Differentiated cells can continue to divide by mitosis. ....   **(9 marks)**

3   Explain why adult stem cells occur in skin and bone marrow.

   ....

   ....

   ....

   ....   **(2 marks)**

4   With reference to specific examples, distinguish between totipotent and pluripotent
   stem cells.

   > You need to show clearly what each term means, how each type of cell is different and give
   > an example of each type of stem cell.

   ....

   ....

   ....

   ....   **(2 marks)**

Medical uses of stem cells

> **Guided**

1 *Describe the production and characteristics of induced pluripotent stem (iPS) cells.

> For 'describe' questions, you do not need to make a judgement or explain how or why, you only need to describe what is requested.

> Your answer should show how iPS cells are made as well as their characteristics.

iPS cells are made from adult cells, such as fibroblasts, that are multipotent

and found in skin and other tissues. The fibroblasts are cultured and four genes,

called Yamanaka factors, are added to the cells. ....

....

....

....

....   **(6 marks)**

> **Guided**

2 (a) Complete the table below to show two benefits and two drawbacks of using embryonic stem cells for medical treatments.

	Considerations of medical uses of embryonic stem cells
Benefits	can produce almost all cell types to grow replacement organs and tissues;
Disadvantages	ethical considerations;

(4 marks)

(b) *Discuss the ethical issues associated with using embryonic stem cells for medical treatments.

> When asked to 'discuss', try to give equal numbers of pros (points for) as cons (points against). You can include valid ethical arguments other than those stated in textbooks but stay objective.

....

....

....

....

....

....

....   **(6 marks)**

Recombinant DNA

1 Identify which function matches which term.

   A   can be used as a vector to carry a novel gene into a cell / organism

   B   used to cut a section of DNA at specific recognition sites

   C   used to join a sugar of one nucleotide to a phosphate of another within a polynucleotide

   D   used to catalyse the formation of complementary DNA (cDNA) using mRNA as a template

   E   unpaired nucleotide bases that can combine with their complementary base pairs

   Restriction endonucleases: ....   Reverse transcriptase: ....

   Sticky ends: ....   Plasmid: ....   DNA ligase: ....   **(5 marks)**

2 Novel genes can be introduced into eukaryotic cells using viruses, liposomes, gene guns and microinjection.

   > This question asks you to apply your knowledge from other topics as well as this one and come up with a response which shows you understand the underlying biological principles.

   (a) Explain why liposomes can pass through the cell surface membrane and the nuclear envelope of a eukaryotic cell.

   ....

   ....

   ....   **(2 marks)**

> Guided >

   (b) Suggest why viruses are not always suitable vectors for gene therapy that needs to be repeated at intervals.

   After the first use of the virus as a vector, the recipient mounts an immune response to the virus. ....

   ....

   ....

   ....   **(3 marks)**

3 The diagram shows how a bacterial plasmid has been cut open using a restriction enzyme. It also shows three novel genes, one of which could be inserted into the plasmid to make recombinant DNA.

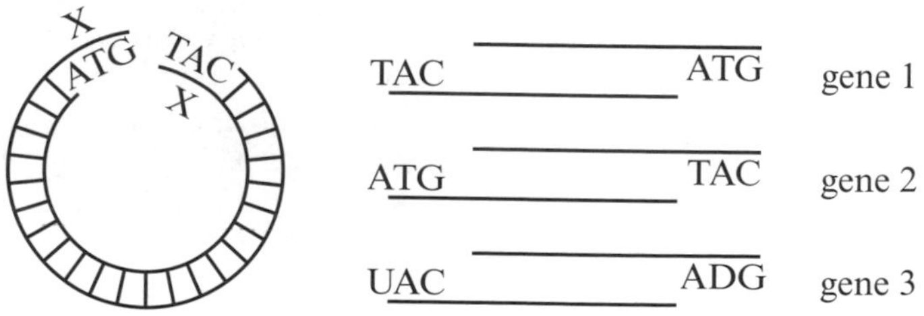

   TAC _____ ATG   gene 1

   ATG _____ TAC   gene 2

   UAC _____ ADG   gene 3

   (a) Name the structures labelled X.

   ....   **(1 mark)**

   (b) Which novel gene could be inserted into this plasmid?

   ....   **(1 mark)**

Identifying recombinant cells

1 Bacterial cells that have been modified by recombinant DNA technology can be identified using

☐ **A** heat shock treatment followed by exposure to calcium chloride

☐ **B** restriction endonuclease enzymes and liposomes

☐ **C** marker genes and replica plating on a medium that contains an antibiotic

☐ **D** embryonic stem cells with genes knocked out

(1 mark)

2 The diagram shows how the gene for human insulin is inserted into the plasmids of the bacterium *E. coli*. *E. coli* bacteria are then mixed with the recombinant plasmids, subjected to heat shock (one minute at 0 °C followed by one minute at 40 °C) in the presence of calcium chloride ions to make their walls more porous, and then plated out onto agar containing ampicillin. Replica plating is then used to transfer bacteria from the ampicillin agar plate to a plate containing tetracycline agar.

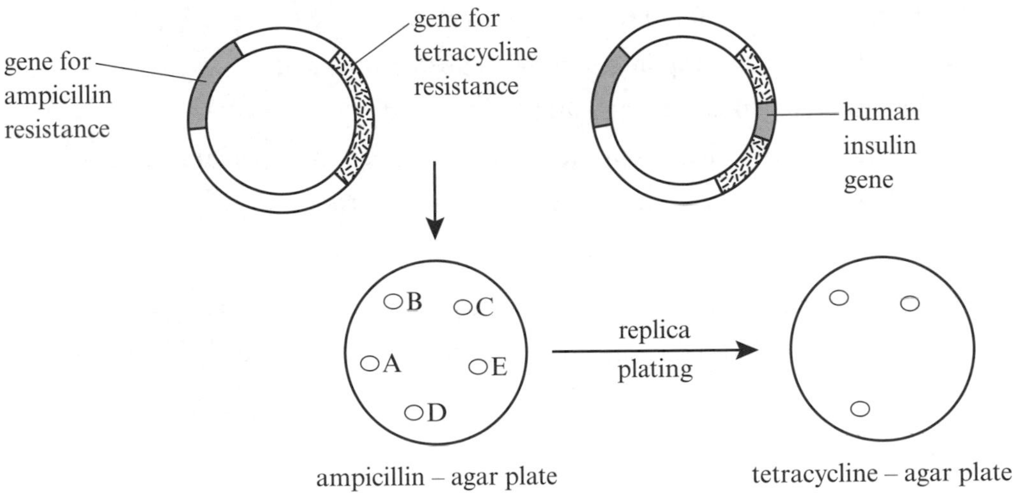

(a) Describe how replica plating is carried out.

....

....

.... **(2 marks)**

(b) Which of the colonies, labelled A, B, C, D, E, contain *E. coli* bacteria that will be able to produce human insulin? Give reasons for your choice.

....

....

....

....

.... **(3 marks)**

(c) Explain why **all** the bacteria in each of the chosen colonies will be able to produce insulin.

....

....

.... **(2 marks)**

Genetic modification of crops

1 Which organism can be used to introduce novel genes into plants?

☐ **A** *E. coli*

☐ **B** *A. tumefaciens*

☐ **C** *T. aquaticus*

☐ **D** *P. infestans*

(1 mark)

> **Guided**

2 The stages of genetic modification of soya plants are given below.

(a) Place the stages in the correct sequence.

A Soya plants are infected with bacteria carrying the modified plasmid.

B The plasmid becomes part of the plant's genome.

C T_1 plasmid is extracted from the bacterium.

D Infected cells develop into a growth called crown gall.

E The desired gene is inserted into the plasmid.

F Single cells are taken from the gall and cultured in a special medium.

G New plants that contain the desired gene are produced.

C .... G

(5 marks)

(b) Explain why every cell in each newly produced genetically modified soya plant contains the desired gene.

> Think about how the cells in the new plant are produced.

....

....

....   (2 marks)

> **Guided**

(c) 'All crop plants are genetically modified by selective breeding and this is a far more hit or miss affair than modern genetic modification'. Discuss this statement.

> For 'discuss' questions, you should consider all aspects of the information, identify issues and explore them and give reasoned explanations.

Selective breeding has caused many genetic changes in crop plants, including changing the numbers of chromosomes. When cross breeding to introduce genes for desired characteristics you cannot be sure which genes are getting into the offspring and sometimes undesirable genes end up in the offspring and some desirable genes may be lost. ....

....

....

....

....

....   (5 marks)

Exam skills

1 FOXP2 (forkhead-box protein) is a protein that, in humans, is encoded by the *FOXP2* gene. It is required for the proper development of speech and language. The gene is on chromosome 7 and is expressed in fetal and adult brain, heart, lung and gut. FOXP2 protein contains a forkhead-box DNA-binding domain. It is a transcription factor that has many different target genes. Mutations to the *FOXP2* gene lead to severe language and speech disorders called developmental verbal dyspraxia, DVD. Individuals with a mutation to one of the alleles of the gene locus suffer with the disorder. They are usually not cognitively impaired but cannot coordinate the movements needed for speech. In some cases, people with DVD have some difficulties with comprehension. The *FOXP2* gene may also be crucial to the normal development of lung and gut tissues. It has been called 'the language gene' but there are other genes involved in language development. *FOXP2* directly regulates many other genes. Different mutations have been identified in the *FOXP2* gene: one causes a change to one or two amino acids that alter the DNA-binding domain of the FOXP2 protein, preventing it from binding to DNA. Knockout mice with only one functional allele of the *FOXP2* gene have reduced vocalisations. Knockout mice with no functional alleles of this gene die very young from inadequate lung development. DNA sampling from Neanderthal bones shows their *FOXP2* gene is similar to that of modern humans.

(a) Explain the sentence: 'It is a transcription factor that has many different target genes.' (line 5).

....

....

.... **(2 marks)**

(b) Explain why a very small change to the amino acid sequence of the FOXP2 protein prevents it from binding to DNA.

....

....

.... **(2 marks)**

(c) What evidence is there in this information that Neanderthal humans could speak and that the *FOXP2* gene is involved in lung development?

....

....

....

.... **(3 marks)**

(d) What type of inheritance pattern is involved in the inheritance of development vocal dyspraxia?

.... **(1 mark)**

Sources of genetic variation

1   Gene or point mutations may involve substitution, insertion or deletion of a base.

(a) Explain how a point mutation may result in variation.

> Point mutation results in a change in a base triplet / codon of the genetic code – see page
> 108 in the Revision Guide to remind yourself what might happen if a codon changes.
> Use the terms base triplet / codon, translation and amino acid in your answer.

....

....

.... **(3 marks)**

Guided

(b) Explain which type of point mutation is least likely to cause a major change in
     the organism.

Point mutations may involve insertion, deletion or substitution. The first two result in a

....

....

.... **(3 marks)**

2   Whole chromosome mutations usually have a major impact on an organism.
    Give one example of a condition in humans caused by a whole chromosome
    mutation and explain its cause.

....

....

.... **(3 marks)**

3   Describe the process of random (independent) assortment during meiosis and
    explain how it contributes to variation.

> Random assortment can occur twice in meiosis: once involving chromosomes and once
> involving chromatids. Remember to state which stage(s) of meiosis are involved.

....

....

....

....

....

.... **(5 marks)**

Understanding genetic terminology

1 Fill in the gaps to complete the following passage using genetic terminology.

The characteristics of organisms are coded for by ....

The .... is the total genetic information of the organism

and includes all the .... (versions of each gene) that the

individual possesses. When the genetic information is expressed, it gives rise to the

characteristics of the organism known as the .... and this

is also influenced by the environment. The body cells of most organisms contain

pairs of chromosomes and are .... This means that these

cells have .... copies of each gene although they may be

.... or .... depending on the

combination of alleles they have.   **(8 marks)**

2 In pea plants, purple flowers is a dominant characteristic and white is recessive.
Suggest how you could find out whether a purple flowering plant is heterozygous or
homozygous.

....

.... **(1 mark)**

3 The human blood group gene has three alleles, I^A, I^B and I^O.
I^A and I^B are codominant and I^O is recessive. What are the
possible genotypes of people with the following blood groups?

> There may be more than one possible genotype for each blood group.

Group A ....

Group AB ....

Group O ....

Group B ....   **(4 marks)**

4 Siamese cats have cream fur with darker fur at the extremities,
such as the face, ears, lower legs and tail. Their coat colour is
a partial form of albinism and is controlled by a single gene.
When they are born, they are cream all over and, although
they do not grow more fur, they slowly develop the darker
extremities. Suggest a reason for this.

> If you cannot explain an observation by only involving genes, then the environment must also play a part in causing that variation.

....

....

.... **(2 marks)**

109

Genetic crosses and pedigrees

Guided   **1**   Using the correct symbols (see page 110 in the Revision Guide), draw a genetic diagram to find the probability of a child with blood group O being born to a heterozygous man with blood group B and a heterozygous woman with blood group A.

> Once you have the parental genotypes, you can fill in the Punnett square. You need to identify the phenotypes of each of the four possible genotypes for the children and add those to Punnett square too.

	Mother	Father
parent phenotypes	blood group A	blood group B
parent genotypes	$I^A I^O$	$I^B I^O$
gamete genotypes	I^A I^O	I^B I^O

.... out of 4 children will have blood group O, so the probability is .... or ....%.

(4 marks)

2   Polydactyly is a genetic condition that results in an individual being born with extra fingers or toes. It is caused by a dominant allele of a single gene. Why can one parent with the polydactyly allele pass on the condition to their child even if the other parent is unaffected?

....

.... **(1 mark)**

3   *Using your knowledge of genetics, explain how the appearance of the condition Down's syndrome is different from the inheritance of genetic conditions such as polydactyly or cystic fibrosis.

....

....

....

....

....

....

....

....

....

.... **(9 marks)**

Dihybrid inheritance

1   A woman has the genotype AABB. What is the genotype of all the eggs in her ovaries?

> As the woman is homozygous for both genes, which are on different chromosomes, all the eggs will have just one genotype.

Genotype of eggs is ....   **(1 mark)**

2   A man has the genotype CcDd. What are the possible genotypes for his sperm?

> As the man is heterozygous for both the genes, which are on different chromosomes, there are four possible genotypes for his sperm.

Possible sperm genotypes are ....   **(4 marks)**

Guided

3   A man with cystic fibrosis (CF) is heterozygous for blood group A. His partner does not have CF and does not carry the allele for cystic fibrosis. She is blood group O.

(a)   What is the genotype of each of these two people?

> Choose a letter to represent the *CFTR (cystic fibrosis transmembrane regulatory)* gene.

> Remember each parent will have two alleles for each gene but their gametes will have just one.

I^A allele for A antigen on erythrocytes, I^O allele for no antigen on erythrocytes.

Genotype of man is ....

Genotype of woman is ....   **(2 marks)**

(b)   Use a genetic diagram to determine the genotypes and phenotypes of any children they may have.

> You need to work out the parent phenotypes, parent genotypes and the possible gamete genotypes.

> As the mother is homozygous for both genes, she will only have one gamete genotype whereas the father will have two.

Phenotypes of children: ....

....   **(5 marks)**

(c)   What is the probability that one of their children will be a symptomless carrier for CF and have blood group A?

....   **(1 mark)**

Autosomal linkage

1   What are autosomes?

     ☐ **A**   chromosomes that do not pair up to form homologous pairs

     ☐ **B**   pairs of chromosomes where each member of the pair has the same genes at the same loci

     ☐ **C**   chromosomes not involved in the inheritance of sex

     ☐ **D**   chromosomes that have duplicated and consist of chromatids   **(1 mark)**

2   Which of the following statements are true? | Remember the difference between alleles and genes. |

    i   Independent assortment only occurs when the genes in question are on different chromosomes.

    ii   Genes on the same chromosome are said to be linked.

    iii   The chromosomes in gametes always contain the same linked alleles as those in the parent cell.

     ☐ **A**   statements i, ii and iii

     ☐ **B**   statements i and ii only

     ☐ **C**   statements ii and iii only

     ☐ **D**   statement iii only   **(1 mark)**

3   The diagram shows pair number 3 of the homologous chromosomes found in a cell, in a testis, at beginning of interphase.

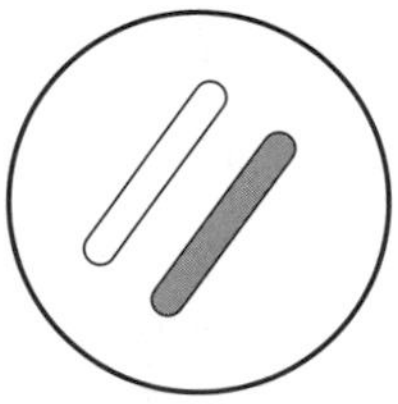

    (a)   Draw a diagram to show the appearance of these chromosomes at the end of interphase and just before prophase I of meiosis.

| When asked to draw a diagram, make sure you include clear and accurate labels and use a ruler where necessary. |

     (3 marks)

    (b)   The diagram shows these chromosomes in some gametes produced in this testis. Use this diagram and your knowledge to explain what recombinant gametes are.

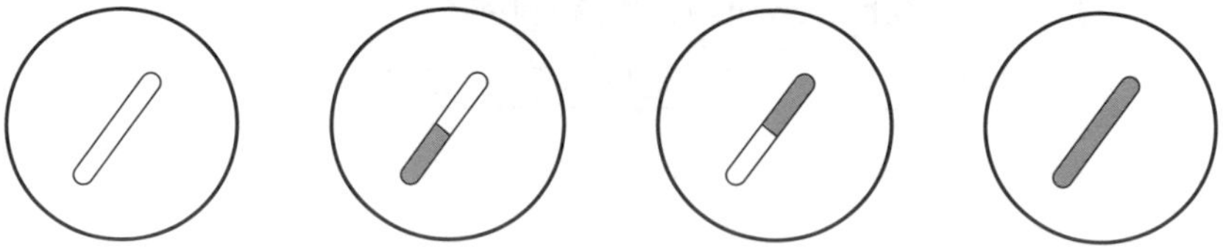

....

....

....   **(2 marks)**

Sex linkage

1   Duchenne muscular dystrophy, DMD, is a disease that occurs more often in boys.
The gene, *dystrophin*, situated on the X chromosome, codes for a very large protein,
dystrophin, involved in contraction of skeletal and heart muscle cells. Dystrophin is
associated with other proteins. It links actin to cytoskeleton threads near the muscle
cell surface membrane. About two-thirds of patients with DMD have a deletion
of one or more exons from their *dystrophin* gene. The *dystrophin* gene is one of the
longest human genes, containing 2.5 megabases (2.5×10^6), making up 0.08% of
the human genome. The primary transcript contains 2500 kilobases and the mature
mRNA, corresponding to 79 exons, measures 14 kilobases.

(a)   Explain why Duchenne muscular dystrophy
is far more common in boys than in girls.

> This is an 'explain' question,
> so it is not enough to state
> your point – you must also
> give reasons or say why you
> have come to that conclusion.

....

....

....   **(2 marks)**

(b)   Explain the following terms:

(i)   exons

....

....   **(1 mark)**

(ii)   human genome

....

....

....   **(2 marks)**

Maths
skills

(c)   State the total number of bases in all the introns
of the *dystrophin* gene. Show your working.

> A gene contains introns and exons.
> If just the exons are left then the
> introns have been removed, so you
> can find out how many bases made
> up the introns by subtraction.

Answer ....   **(2 marks)**

(d)   Draw a genetic diagram to predict the possible genotypes and phenotypes of
offspring produced by a couple, where the woman is a symptomless carrier of
DMD and the man is not affected with DMD.

(5 marks)

Selection pressure and genetic drift

> **Guided**

1 *The grove snail, *Cepaea nemoralis,* exhibits a high degree of polymorphism in its shell colouration. The background colour of the shell may be yellow, pink, chestnut or dark brown. The shell may or may not have up to five dark bands. The colour and banding are determined by genes, each of which has several different alleles. Yellow shell is recessive to pink shell and both yellow and pink are recessive to brown shell. The unbanded shell is dominant to banded. It is common and widespread throughout Western Europe. Pale and unbanded specimens are more prevalent in grassy habitats, whereas darker and banded forms are found in larger numbers in woodland and hedgerow habitats. Grove snails breed from April to October. Eggs are laid in a depression made in the soil that is then covered and each brood produces about 23 offspring. Birds, such as the song thrush, are predators of the grove snail. Suggest how natural selection could account for the distribution of the different forms of *Cepaea nemoralis,* in woodland and hedgerow areas, compared to open grassland habitats.

> This question asks you to apply your knowledge about natural selection to a possibly new situation. There are at least five more points to make as you need to explain clearly and thoroughly, using correct terminology and A level biology, how natural selection operates.

Snails are killed and eaten by birds such as thrushes. Predation is a selection pressure. Snails that are better camouflaged will be less likely to be seen by the birds and eaten. They will survive longer and will ....

....

....

....

.... **(6 marks)**

2 Pingelap is a small atoll (island) in the Pacific Ocean. A significant proportion of the very small population has a rare form of total colour blindness called achromatopsia. They have no functioning cone cells in their eye retinas, only rod cells. They cannot perceive colours. In 1775, a typhoon swept through Pingelap and only 20 people survived. Before this typhoon there were no cases of achromatopsia; by four generations after the typhoon, 2.7% of the population was affected. Today, 10% of the 250 population has the disorder and 30% are symptomless carriers. In the USA only 0.003% of the population has this disorder.

Maths skills

(a) Calculate how many people today on Pingelap have achromatopsia.

Answer .... **(1 mark)**

(b) Suggest why the incidence of achromatopsia on Pingelap is higher than in the United States.

....

....

....

.... **(3 marks)**

Hardy–Weinberg equilibrium

1 What does the Hardy–Weinberg equation predict?

☐ **A** The difference between observed and expected results is due to chance.

☐ **B** Natural selection and genetic drift can contribute to changing allele frequencies and to evolution.

☐ **C** The frequency of alleles within a population will stay the same unless factors act to affect them.

☐ **D** The ratio of phenotypes in the offspring from a dihybrid cross is 9 : 3 : 3 : 1.   **(1 mark)**

2 What **five** assumptions are made to assume there will be no changes to allele frequencies within a population?

....

....

....

....

....   **(5 marks)**

> **Guided**
> **Maths skills**

3 In a population of 2000 humans, 400 of them cannot smell honeysuckle flowers. This type of anosmia (inability to smell something) is a recessive genetic condition. Calculate the number in this population who are homozygous and able to smell honeysuckle flowers and heterozygous and able to smell honeysuckle flowers.

The 400 people who cannot smell honeysuckle flowers must be homozygous recessive, so we know that $p^2 = 400 \div 2000 = 0.2$.

> Now you know the value of p^2, you can calculate the value of p. Once you know the value of p you can then calculate the value of q using the equation $p + q = 1$. You can then calculate the value of q^2 and use the other equation to find the value of $2pq$.

The number of people in the 2000 population who are homozygous and can smell

honeysuckle is ....

....

So the frequency of heterozygotes is ....

....   **(4 marks)**

Exam skills

1 The fossil record indicates that sexual reproduction began on Earth about 1.2 billion years ago; all sexually reproducing eukaryotic organisms derive from a single-celled common ancestor. Sexually reproducing individuals can only pass 50% of their genetic material to their offspring, which could be seen as a disadvantage. However, sexual reproduction increases genetic variation within a population, which is an advantage.

(a) Explain why organisms that reproduce sexually only pass half of their genetic material to their offspring.

....

....

.... **(3 marks)**

(b) Describe how sexual reproduction contributes to genetic variation.

....

....

....

.... **(5 marks)**

2 In pet rats there are many different coat colours that have been produced by selective breeding. At one gene locus, the allele C leads to agouti coloured fur, the allele c leads to white fur and the allele c^h leads to Himalayan fur – white with a pale brown smudge at the nose and tail. C is dominant, c is recessive and c^h is dominant to c but recessive to C. Himalayan rats have the genotype cc^h. A rat with genotype c^hc^h is described as Siamese. It has cream fur with dark points at nose, feet and tail. Siamese rats are cream all over when first born and the dark points develop a little later. A rat breeder has two agouti rats. He crosses them and obtains several litters. Altogether in the F_1 generation they produce 15 agouti rat kittens and five Himalayan rat kittens. When one of the F_1 agouti rats is mated with a Himalayan rat their offspring are six agouti, three Himalayan and three Siamese.

(a) Draw genetic diagrams:

(i) to show the genotypes of the first set of parent rats and their offspring

(1 mark)

(ii) to explain the result of crossing the two F^1 rats.

(4 marks)

(b) A rat breeder has an agouti male rat. He does not know whether it is homozygous or heterozygous. How could he find out its genotype?

....

....

.... **(4 marks)**

Homeostasis

1 Which type of receptor would detect changes in pH?

☐ **A** baroreceptor

☐ **B** thermoreceptor

☐ **C** osmoreceptor

☐ **D** chemoreceptor

> Think about the prefixes of these words and what they mean. For example 'osmo' in the word osmosis relates to water molecules

(1 mark)

2 Compare and contrast the effect of pH and lower temperatures on the rate of enzyme reactions.

> Low temperatures do not affect the shape of the active site but they do have an effect on the kinetic energy of the enzyme. What effect does pH have on the active site?

> Make sure you use language from the stem of the question and link your answer to the rate of reaction.

....

....

....

....

....

(4 marks)

> **Guided**

3 During a fever, the core temperature set point is raised. This benefits the immune system and inhibits microbial replication. Explain, with reference to the graph, why it is important for our body cells that this set point is not reset too high.

> Remember that not all enzymes will be affected straightaway. The number of enzymes that have been denatured will increase as temperatures increase.

> Aim to make two further points in your answer.

High temperatures can result in the disruption of the tertiary structure of the globular protein and a change in shape of the active site. The active site will no longer be complementary to its substrate and the enzyme is denatured.

....

....

....

....

(3 marks)

Hormones in mammals

1   Hormone releasing cells usually release two types of hormones. Which organelle might be found in greater quantities in these cells?

   ☐ **A**   smooth endoplasmic reticulum

   ☐ **B**   nucleus

   ☐ **C**   large permanent vacuole

   ☐ **D**   centrioles

> Animal cells do not have large permanent vacuoles.

(1 mark)

2   Explain why steroid hormones do not need to bind to receptors on the cell surface membrane.

> Remember hormones are not always proteins.

> This is an 'explain' question, so it is not enough to state your point – you must also give reasons or say why you have come to that conclusion.

....

....

....

.... **(2 marks)**

Guided   3   Explain the function of transcription factors.

> Aim to make one further point.

Transcription factors are molecules that bind to DNA. ....

.... **(1 mark)**

4   Hormones are large molecules that are packaged into vesicles by a secreting cell. Describe how these hormones can leave the cell.

> This question is asking you to use your knowledge from Topic 4.

> For 'describe' questions, you do not need to make a judgement or explain how or why, you only need to describe what is requested.

....

....

....

.... **(2 marks)**

Chemical control in plants

1   Which of these responses is not a result of the production of gibberellins?

    ☐ **A**   elongate stem cells

    ☐ **B**   higher production of amylase enzyme

    ☐ **C**   promote lateral bud growth

    ☐ **D**   stimulate fruit development   **(1 mark)**

Maths skills

2   An investigation was carried out into the phototropic response of plant shoots. Shoot tips were removed and had a plastic divider inserted and were placed on blocks of agar jelly. One was placed in the dark and one was placed in the light. The diagram shows the distribution of auxin in arbitrary units on either side of the plastic divide after three hours.

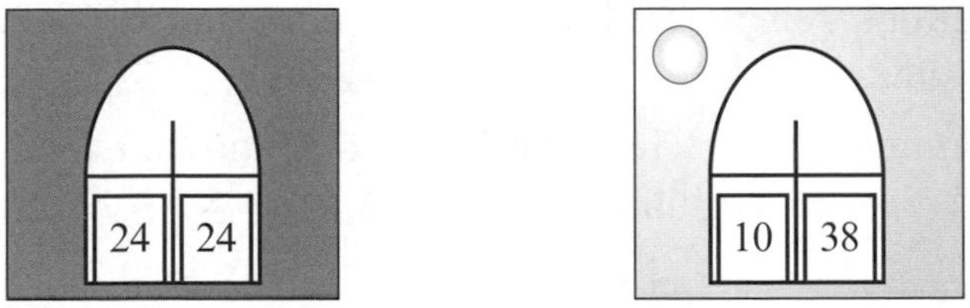

Analyse the data to explain the difference in the distribution of the plant hormone in these two agar blocks and the effect it would have on stem cell growth.

> The word 'analyse' means that you need to comment on the data and then relate your comments to the situation being discussed to make a judgement. If you are given data, it is usually a good idea to use them in your answer!

....

....

....

....

....

.... **(4 marks)**

Guided

3   Explain the changes that would be seen in plant growth if a herbivore ate the tips of a plant shoot.

> Aim to make at least one more point here.

Auxins are involved in apical dominance and prevent lateral bud growth. Removal of the tips would result in no auxin being produced.

....

....

.... **(3 marks)**

Gibberellin and amylase

1 What product is formed when amylase hydrolyses starch?

    ☐ **A** α-glucose

    ☐ **B** β-glucose

    ☐ **C** galactose

    ☐ **D** maltose   **(1 mark)**

2 Black walnut trees produce a chemical called juglone, which can affect grain germination and plant growth in soil near to the tree by reducing the activity of α-amylase. An investigation was carried out into the effect of juglone on the percentage of embryoless barley grains that would germinate over four days.

- One hundred grains were watered with 25 cm³ of distilled water.

- One hundred grains were watered with 25 cm³ of a solution containing juglone and distilled water.

- One hundred grains were watered with 25 cm³ of a solution containing juglone, gibberellic acid (GA) and distilled water.

The results are shown in the graph.

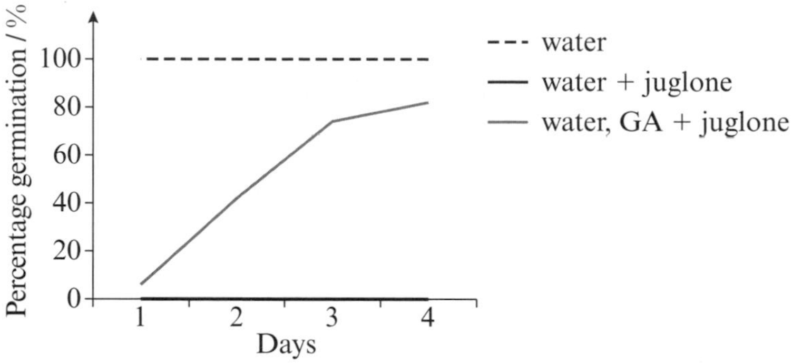

> Make sure you use information given in the question stem and refer to α-amylase.

(a) Analyse the data to explain the effect of juglone on the germination of embryoless seeds.

....

....

....

....

.... **(4 marks)**

> **Guided**

(b) Explain the results for the seeds watered with the solution containing both juglone and gibberellic acid.

> Aim to make one more point here.

80% of seeds watered with both juglone and gibberellic acid germinated after four days. This is because gibberellin from the solution would stimulate the aleurone layer in the seeds to produce α-amylase. The presence of juglone would reduce the activity of the amylase but ....

....

....

.... **(5 marks)**

Phytochrome and photoperiodism

1 Phytochromes are photoreceptors found in many plants.

(a) The diagram shows the interconversion of inactive phytochrome (P_r / P_{660}) and active phytochrome (P_{fr} / P_{730}). State one way in which the active form of phytochrome can be converted back to the inactive form, other than by exposing it to far red light.

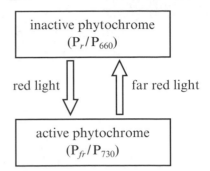

inactive phytochrome
(P_r/P_{660})

red light   far red light

active phytochrome
(P_{fr}/P_{730})

.... **(1 mark)**

(b) A study was carried out to investigate the effect of red light and far red light on sunflower plants. One group of sunflower seedlings, group A, was grown under a lamp that emitted red light and far red light of the same intensity. Another group of sunflower seedlings, group B, was grown in the same way, except that the lamp emitted a lower intensity of red light but the intensity of far red light was unchanged. When the plants were fully grown, the mean dry mass of the flowers produced and the mean length of the plant stems were recorded. This study was repeated using new groups of sunflower seedlings. The results are shown in the table.

Study	Mean dry mass of the flowers / g		Mean stem length / cm	
	Group A	Group B	Group A	Group B
original	58	45	125	148
repeat	43	38	124	142

The light conditions experienced by group B were similar to those found near ground level in woodland. Using the mean stem lengths shown in the table, explain the importance of these light conditions for a young seedling in the woodland.

> Remember that, in a woodland, plants will be trying to grow taller quickly to maximise the levels of light energy they absorb.

> There are three marks available here but be aware that you may need to make more than three points.

....

....

....

....

....

.... **(3 marks)**

The nervous system

1   Complete the table to show three differences in structure between a sensory and motor neurone.

> Comparative statements should be on the same row.

Sensory neurone	Motor neurone

(3 marks)

2   Describe one way in which the **function** of a motor neurone is different from that of a sensory neurone.

> For 'describe' questions, you do not need to make a judgement or explain how or why, you only need to describe what is requested.

....

.... **(1 mark)**

Guided

Maths skills

3   Explain the effect of myelination on the speed of electrical impulse conduction by using the information in the table to help you.

Type of neurone	Diameter of axon / μm	Speed of electrical impulse conduction / ms^{-1}
myelinated	4	34
myelinated	39	47
unmyelinated	3	1
unmyelinated	31	13
unmyelinated	87	36

> Aim to make a further two points in your answer.

The electrical impulses are conducted faster when there is myelination. A greater axon diameter is needed to achieve the same conduction speed when the neurone is unmyelinated. ....

....

....

....

.... **(5 marks)**

The central nervous system

1 The diagram shows a section through a human brain.

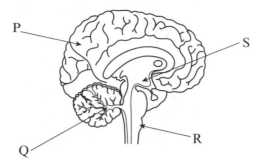

> **Guided**

(a) Complete the table to show the correct name and function for each region of the brain.

Region of brain	Name	Function
P		initiates movement
Q		
R		
S		

(4 marks)

(b) During exercise, chemoreceptors detect a decrease in pH due to increased carbon dioxide. State where nerve impulses are being sent as a result of this.

☐ **A** cerebral hemisphere

☐ **B** cerebellum

☐ **C** medulla oblongata

☐ **D** hypothalamus

(1 mark)

2 Complete the boxes to show the stages in a reflex pathway starting with 'stimulus'.

Think about which neurones are involved in a reflex.

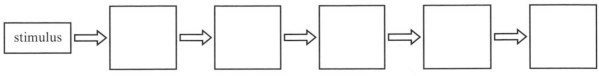

(3 marks)

Resting potential

1 The diagram shows an electron micrograph image of a nerve cell. The actual diameter of the cell as indicated by the arrow is 4 μm.

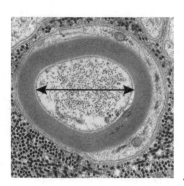

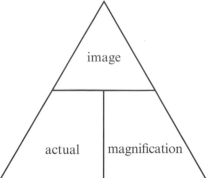

Calculate the magnification of the nerve cell.

Make sure you convert the image size into the same units as the actual size.

Magnification .... **(2 marks)**

2 The graph shows the changes in potential difference across the membrane of a neurone after stimulation. Significant changes are marked by A, B, C, D, E and F.

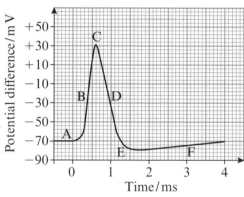

The table describes three of the stages A–F shown in the graph.

(a) Place a tick (✔) in the box below the letter that correctly links the description to one of the labels on the graph.

Description	A	B	C	D	E	F
stage when the concentration of positive ions is greatest inside the axon						
stage when hyperpolarisation first occurs						
stage showing the resting potential						

(3 marks)

(b) Describe the events occurring between E and F which result in re-establishing a potential difference of −70 mV across the membrane.

Remember sodium-potassium pumps require energy.

....

....

.... **(3 marks)**

Action potential

1   Which ions move into the axon at the start of depolarisation to cause an action potential?

   ☐ **A**   calcium

   ☐ **B**   chloride

   ☐ **C**   potassium

   ☐ **D**   sodium   **(1 mark)**

2

   (a)   Which letters on the diagram correspond to the following stages?

   > Note that some letters will be used more than once.

   (i)   repolarisation ....   **(1 mark)**

   (ii)   hyperpolarisation ....   **(1 mark)**

   (iii)   depolarisation ....   **(1 mark)**

   (iv)   resting potential ....   **(1 mark)**

   (v) K^+ gates are open ....   > This is in repolarisation.   **(1 mark)**

   (vi) K^+ gates are closed ....   **(1 mark)**

   (vii) Na^+ gates are closed ....   **(1 mark)**

   (b)   Explain why depolarisation will not always result in an action potential.

   ....

   ....

   ....

   > This is an 'Explain' question, so it is not enough to state your point – you must also say why you have come to that conclusion.

                                                                                          (2 marks)

125

Propagation of an action potential

1 The graph shows how an increase in neurone diameter can affect the speed of conduction.
Using the information in the graph, describe the effect of increasing the diameter of the neurone on the speed of conduction.

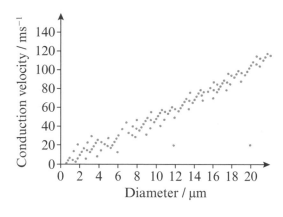

....

....

.... **(2 marks)**

Guided 2 *Non-myelinated neurones have voltage-dependent sodium gates along their entire length. Myelinated neurones only have voltage-dependent sodium gates in gaps between Schwann cells. Explain why there is a difference in the speed of transmission of an action potential along these two types of neurone.

> Aim to make three further points in your answer.

An action potential will pass along a myelinated neurone faster than a non-myelinated neurone because the voltage-dependent sodium gates are only located at the nodes of Ranvier and are therefore further apart in a myelinated neurone. ....

....

....

....

....

....

....

.... **(6 marks)**

Synapses

1 Which ions enter the presynaptic membrane when an impulse arrives at a synapse?

☐ **A** calcium

☐ **B** chloride

☐ **C** potassium

☐ **D** sodium

(1 mark)

> **Guided**

2 Glutamate is a neurotransmitter which can be found in the brain. Explain the events that occur at a synapse after a neurotransmitter, such as glutamate, has been released.

Do not forget to explain what happens to the neurotransmitter afterwards.

Aim to make a further two points.

The neurotransmitter diffuses across the synapse and binds to receptors on the postsynaptic membrane. This stimulates sodium-gated channels to open. ....

....

....

....

....

....   **(5 marks)**

Maths skills

3 The graph shows the changes in potential difference in the postsynaptic membrane. What is the maximum change in potential difference during depolarisation in the postsynaptic membrane?

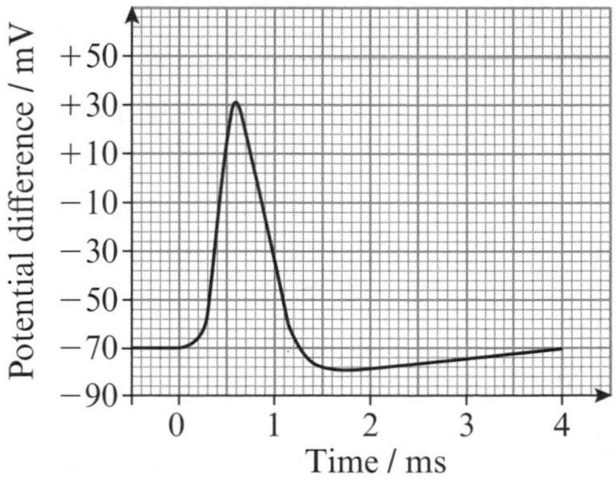

Remember to state the correct units if they are not already provided on the answer line.

....   **(1 mark)**

Had a go ☐ Nearly there ☐ Nailed it! ☐

Drugs and the nervous system

1 Which of these drugs can be used as an anaesthetic?

☐ **A** acetylcholine

☐ **B** cobra venom

☐ **C** lidocaine

☐ **D** nicotine **(1 mark)**

Guided

2 Atropine is a chemical molecule which has a similar shape to the neurotransmitter acetylcholine. If it is present in a synapse it can prevent action potentials occurring in a postsynaptic neurone. Explain how the presence of atropine can result in the inhibition of action potentials.

Think about what you know about the effects of cobra venom.

Atropine can bind to the acetylcholine receptors on the postsynaptic membrane. This blocks the receptor and will prevent acetylcholine from binding.

....

....

....

....

.... **(5 marks)**

Maths skills

3 An investigation was carried out into the effects of nicotine on blood pressure. Analyse the data and explain the effect of nicotine on systolic blood pressure.

....

....

....

....

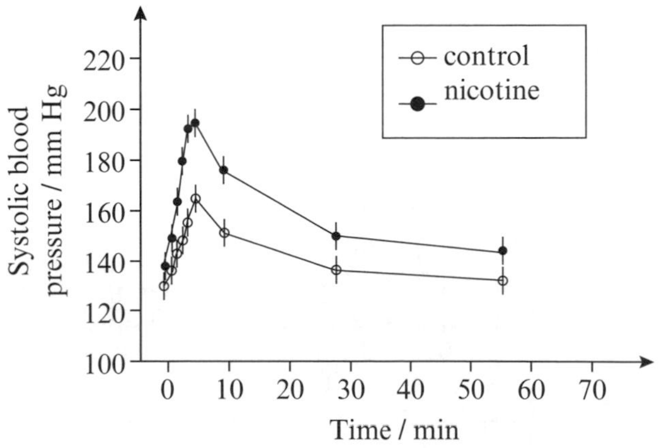

.... **(4 marks)**

Detection of light by mammals

1 The diagram shows the position of some of the cells in the retina of the eye.

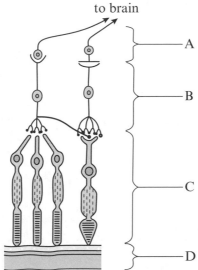

to brain

A
B
C
D

(a) In which part of the diagram, A, B, C or D, would you find the neurones of the optic nerve?

> Diagrams can sometimes be different from what you have seen before. Look at the orientation of the rods and cones.

Answer .... **(1 mark)**

(b) Draw an arrowed line on the diagram showing the direction of light as it hits the retina and state which one of the following choices is correct.

☐ **A** bottom to top

☐ **B** left to right

☐ **C** right to left

☐ **D** top to bottom **(1 mark)**

Guided ▷ 2 *Describe how light causes a change in the voltage across the surface membrane of a rod cell and can lead to the formation of an action potential in an optic neurone.

> Aim to provide three more points in your answer.

Light is absorbed by rhodopsin and the retinal part of rhodopsin changes shape from the *cis* isomer into the *trans* isomer, which cannot bind to the opsin. The opsin binds to the cell surface membrane. ....

....

....

....

....

....

....

.... **(6 marks)**

Had a go ☐ Nearly there ☐ Nailed it! ☐

Control of heart rate in mammals

1 Draw a cross (X) on the area of the brain responsible for controlling heart rate.

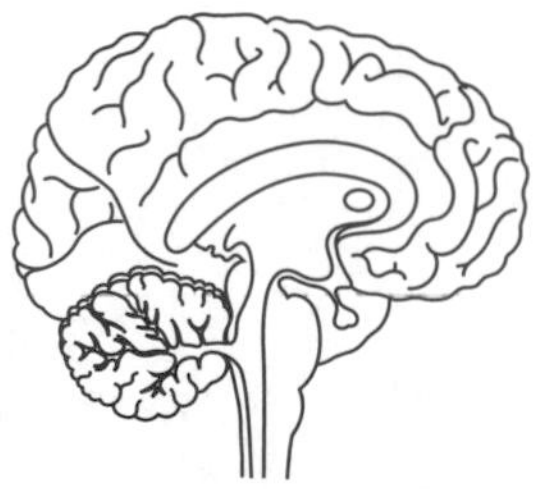

(1 mark)

2 Exercise can result in the increase in blood carbon dioxide levels. Explain what causes this increase.

> Exercise involves muscle contraction and relaxation which requires ATP.

....

....

.... **(2 marks)**

3 Which receptors detect the increase in blood carbon dioxide levels?

☐ **A** baroreceptors ☐ **C** osmoreceptors

☐ **B** chemoreceptors ☐ **D** thermoreceptors **(1 mark)**

4 *The effect of exercise on the heart rate of a trained athlete was investigated. Use the data in the graph to help you explain how the autonomic nervous system is involved in changing the heart rate during and after exercise.

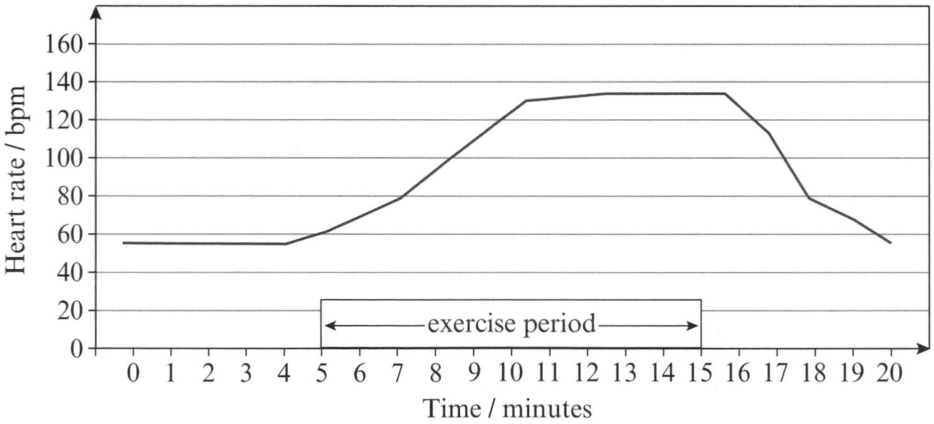

> Your answer needs to be clear and logically structured and refer throughout to the data you have been given. The question also wants you to explain how the heart rate is decreased.

....

....

....

....

....

....

....

....

.... **(6 marks)**

Structure of the mammalian kidney

Guided ⟩

1 The diagram shows a mammalian kidney.

Refer to the diagram, then put ticks (✓) in the appropriate boxes in the table. One tick has been inserted in the table to help you.

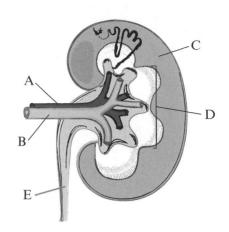

	A	B	C	D	E
Where the glomerulus is located					
Contains podocytes					
Where the collecting ducts are located					
Contains the lowest concentrations of urea		✓			
Affected by ADH					

(4 marks)

2 Amino acids are broken down in the liver by removing an amine group.

(a) Draw a diagram to show the generalised structure of an amino acid and label the amine group.

> When asked to draw a diagram, make sure you include clear and accurate labels and use a ruler where necessary.

(2 marks)

(b) Name two molecules that can be produced from this amine group.

....

.... **(2 marks)**

(c) The rest of the amino acid can be turned into glucose. Glucose molecules can be turned into a polysaccharide for storage in cells. Name this polysaccharide and explain how it is formed from glucose.

> This is a synoptic question and is testing Topic 1 knowledge.

....

....

....

....

.... **(3 marks)**

131

Kidney function

Maths skills

1 A mammalian kidney receives 1350 cm^3 of blood every minute, of which 775 cm^3 is plasma. During ultrafiltration, 138 cm^3 of filtrate passes into the proximal convoluted tubule. Calculate the percentage of plasma that passes into the proximal convoluted tubule.

> If you are not provided with a unit on the answer line then you are expected to give one to gain full marks. Here, the question asks for a percentage so the unit is %.

> Remember you need to divide what has passed through into the proximal convoluted tubule by the volume of plasma and then turn the answer into a percentage.

Answer .... **(2 marks)**

Guided

2 * The mammalian kidney is involved in the excretion or reabsorption of various molecules. The bar chart shows the concentration of various molecules in the blood plasma, the glomerular filtrate and urine of a mammal. Explain how the changes in concentration of the substances are brought about by the kidney.

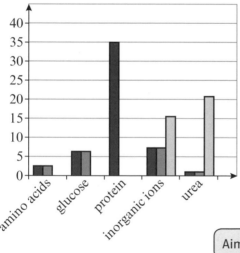

■ concentration in blood plasma (g / l)

■ concentration in glomerular filtrate (g / l)

☐ concentration in urine leaving the collecting duct (g / l)

> Remember to state the locations of any substance reabsorption you mention and refer back to the data to clarify the points you make.

> This is a levels-based question. You need to write a clear, well-structured answer that uses key scientific terminology throughout. To do well, you must support your points with appropriate reference to the data supplied.

> Aim to make at least three more points in your answer.

When ultrafiltration occurs in the glomerulus, the filtrate passes into the lumen of the Bowman's capsule. Blood cells and large proteins are too big to pass through the gaps in the capillaries ....

....

....

....

....

....

....

(6 marks)

Osmoregulation

1 Which receptors detect the increase in blood water levels?

☐ **A** baroreceptors

☐ **B** chemoreceptors

☐ **C** osmoreceptors

☐ **D** thermoreceptors **(1 mark)**

2 (a) State the part of a nephron which is affected by ADH.

.... **(1 mark)**

(b) Explain the effect that increased ADH levels would have on the concentration of the filtrate.

....

....

.... **(3 marks)**

3 The diagram shows nephrons from three mammals.

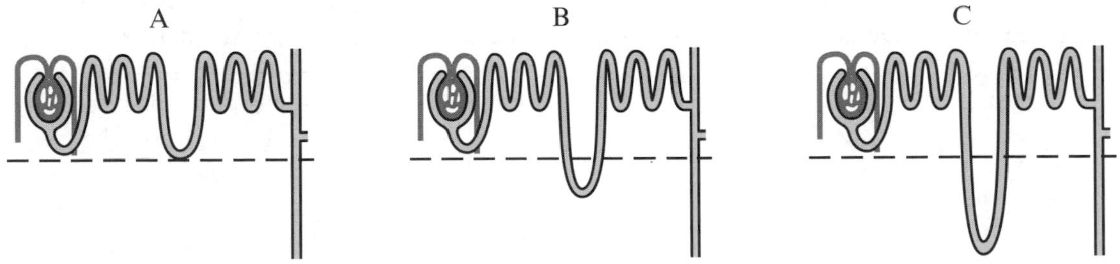

(a) State which of the nephrons, A, B or C, would create a higher concentration of urine.

.... **(1 mark)**

(b) Explain your answer to (a).

....

....

....

....

.... **(3 marks)**

Thermoregulation

1 Some animals demonstrate physiological responses to maintain their core body temperatures within narrow limits. Bodily responses have different effects on heat energy levels in the body. Complete the table using ticks (✓) to indicate the most appropriate answers.

> Make sure your ticks are unambigous. ✗ will be marked wrong.

	Sweating	Vasodilation	Vasoconstriction	Shivering	Contraction of hair erector muscles	Relaxation of hair erector muscles
Increases heat energy loss						
Reduces heat energy loss						
Increases heat energy production						

(3 marks)

2 *A person sitting in a sauna will have an initial increase in core body temperature. This is shown in the graph. Using the information in the graph, explain how this increase is detected and how it is brought back to within normal parameters.

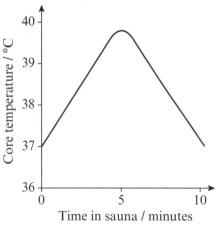

> This is a levels-based question. You need to write a clear, well-structured answer that uses key scientific terminology throughout. To access higher than two marks you must support your points with appropriate reference to the data supplied.

....

....

....

....

....

....

....

....

(6 marks)

Exam skills

1 The graph shows how the diameter of a neurone can affect the speed of an electrical impulse.

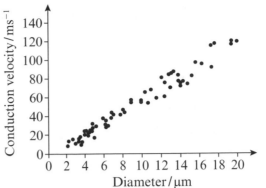

(a) Draw an appropriate line of best fit on this graph.

(1 mark)

(b) Describe how the diameter of a neurone can affect the speed of an electrical impulse.

....

....

....   **(2 marks)**

(c) Multiple sclerosis affects the myelination of neurones in the central nervous system. Plot a point on the graph for a neurone with a diameter of 12 μm which is affected by multiple sclerosis.

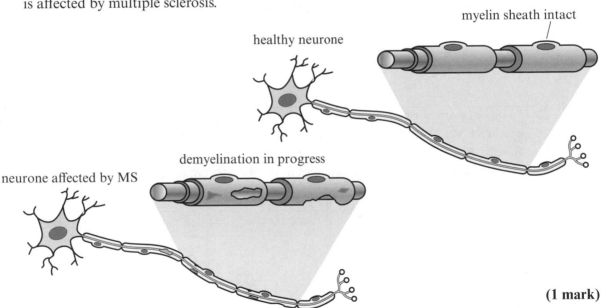

healthy neurone

myelin sheath intact

demyelination in progress

neurone affected by MS

(1 mark)

2 Compare and contrast the production, transport and action of plant and animal hormones.

> The question is asking about three aspects of plant and animal hormones. Make sure you state similarities and / or differences, where appropriate, for each aspect.

....

....

....

....

....

....   **(4 marks)**

Ecosystems

1 Complete the passage using the most appropriate terms in the blank spaces.

Organisms living in the same place or .... interact continuously

with each other and their environment. Each individual is a member of a group, the

.... which includes all the organisms of the same species that breed

together in that place. All the organisms of different species living in that area form a group

called the .... All the organisms and the non-living factors interact

to form a unit known as the .... Organisms which photosynthesise

form the first feeding or .... level in a food chain and are described as

.... because they can make their own food. Photosynthetic organisms

are a source of energy and nutrients for the animals in the food chain. These animals are

described as ....

> Make sure your answer refers to the meaning of both terms.

(7 marks)

2 Identify the difference between a habitat and a niche.

....

.... **(1 mark)**

Guided

3 *The diagram shows a pyramid of biomass and a pyramid of energy for organisms in the English Channel. Assess the diagram and then complete the paragraph to explain why the pyramid of biomass in this ecosystem is inverted but the energy pyramid is not.

> Some types of organisms complete their life cycles much faster than others.

> A pyramid of biomass represents a 'snapshot in time' and the pyramid of energy takes the productivity over a whole year into account.

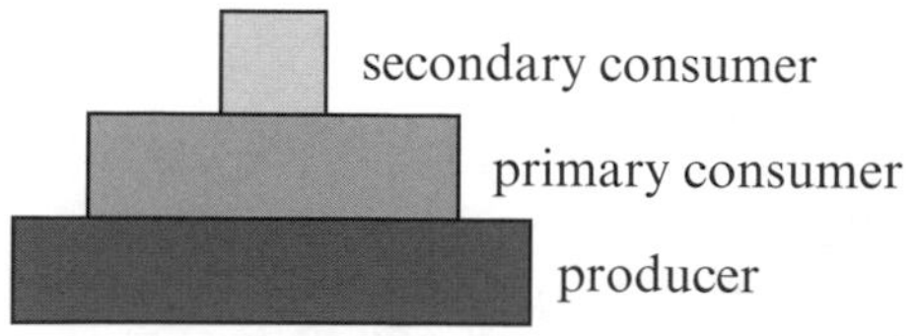

secondary consumer
primary consumer
producer

pyramid of biomass measured in kg / ha⁻¹

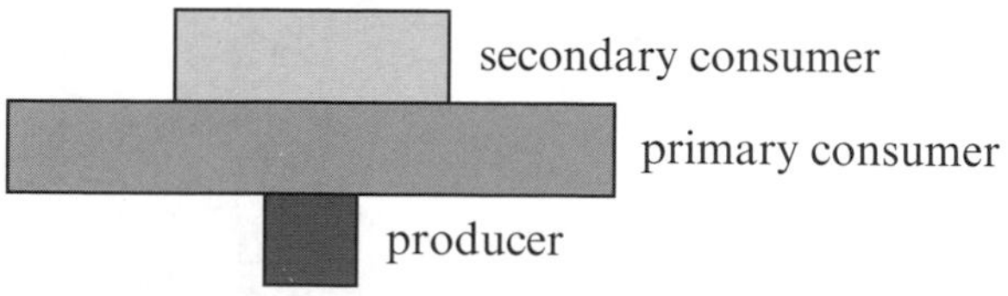

secondary consumer
primary consumer
producer

pyramid of energy measured in kJ ha⁻¹ / yr⁻¹

The pyramid of biomass shows the mass of all the organisms collected at ....

whereas a pyramid of energy is a more accurate representation as ....

In this ecosystem, the producers reproduce .... than the

consumers so ....

....

....

.... **(6 marks)**

Ecological techniques

1   When counting the number of individual organisms with a quadrat, why is it important to do a number of readings rather than just one?

> An accurate measurement is one that is close to the 'true value'.
> There may be more than one statement that applies.

> You know it is not E as the comparison will also depend on how the data at the other site are collected.

☐ **A** It improves the accuracy and reliability of the estimated abundance.

☐ **B** It is more likely to give an abundance that is representative of the area and increases the validity of the data collected.

☐ **C** It means that the abundance can be accurately and precisely measured.

☐ **D** It means that a measure of the variability of the data can be calculated such as the standard deviation.

☐ **E** It increases the validity of any comparisons made with data from other sites.   **(1 mark)**

> **Guided**

2   Periwinkles are gastropod molluscs found on rocky shores feeding on algae. Some species of periwinkle are more tolerant of desiccation than others and therefore different species of periwinkle are found at different heights on the shore. Describe how you would collect and display data to assess the impact of climate change on the distribution of periwinkle species on a rocky shore.

> Periwinkles move very slowly so their abundance can be measured by counting the number of individuals in a quadrat.

(a) Explain how to do an interrupted belt transect.

....
....
....
....
....   **(3 marks)**

(b) Explain how to ensure the data are accurate and reliable and suggest an appropriate statistical test to be done.

....
....   **(2 marks)**

(c) Explain how you will find out whether any change is taking place.

....
....   **(2 marks)**

(d) Is there any other information you usefully could collect?

....   **(1 mark)**

Sampling methods

Practical skills

1 A student set out to investigate the population of slugs in a large flower bed. She obtained the following data during a preliminary investigation.

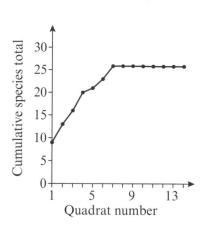 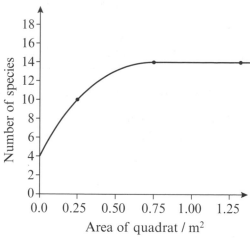

(a) Suggest how these data might have been collected and explain how the method works.

....

....

.... **(2 marks)**

(b) How will this information help in the design of her full investigation? Explain your answer.

....

....

.... **(2 marks)**

(c) Suggest one way in which the student would need to alter the investigation if she were estimating the population of spiders in the flower bed.

.... **(1 mark)**

2 Which of the following statistical measures would give the student the best understanding of the spread of the data she collected in the above investigation?

☐ **A** chi squared test

☐ **B** mean

☐ **C** range

☐ **D** standard deviation

☐ **E** student's *t*-test   **(1 mark)**

Statistical testing

1   Students studying the sand dune systems at Orielton, Pembrokeshire, noticed that the marram grass at the top of the dunes grew taller than it did lower down. They decided to investigate this and collected the following data by working along a line transect that ran from 0 m at the bottom of the dune to the top of the dune.

Height up dune / m	Height of marram grass / mm	Dune height rank	Grass height rank	D	D^2
0	309	1	1	0	0
1	804	2	3	+1	
2	898				
3	867				
4	746				
5	889				
6	945	7	7.5	+0.5	0.25
7	945	8			
8	976				
9	967	10	9	−1	1
10	1004				
			sum of D^2 (ΣD^2)		

> If ranks are shared, then share the two ranks that the data occupy.

(a)   Complete the calculations in the table and then calculate r_s using the formula below.

$$r_s = 1 - \frac{6(\Sigma D^2)}{n(n^2 - 1)}$$

where: n = number of pairs of data;

D = difference between the ranks; r_s = Spearman rank correlation coefficient

> Remember to calculate $n^2 - 1$ before multiplying by n.

Answer: r_s is ....   **(1 mark)**

(b)

n	0.01	0.05	0.02	0.01
5	0.900			−
6	0.829	0.886	0.943	−
7	0.714	0.786	0.893	−
8	0.643	0.738	0.833	0.881
9	0.600	0.683	0.783	0.833
10	0.564	0.648	0.745	0.794
11	0.523	0.623	0.736	0.818
12	0.497	0.591	0.703	0.780
13	0.475	0.566	0.673	0.745
14	0.457	0.545	0.646	0.716
15	0.441	0.525	0.623	0.689

Now use the table of critical values to find out whether the null hypothesis should be accepted or rejected and explain how you came to your conclusions.

> In your answer, you should refer to the critical value at $p = 0.05$ and how it relates to r_s.
>
> First decide which row of the table to use, then decide on the column you need.

....

....

....

....   **(4 marks)**

Energy transfer through ecosystems

Maths skills

1 The table shows the flow of energy through three contrasting ecosystems. Calculate the missing figures and use them to complete the table.

> If you are unsure how to do a calculation, use the numbers in the completed columns to work out what to do.

	Mature rain forest / kJ / m / yr	Field of alfalfa / kJ / m / yr	Young pine forest in UK / kJ / m / yr
GPP (rate at which light energy is converted to chemoical energy during photosynthesis)	188 000	102 000	51 000
Respiration	134 000		20 000
NPP (GPP minus energy released during respiration)	54 000	64 000	

(2 marks)

2 How much energy is available to be passed on to primary consumers in the rain forest?

> Only energy that is used to produce more biomass during growth can be passed on.

.... **(1 mark)**

Maths skills

3 In the young pine forest, 4 000 000 kJ of light energy reaches the producers. Calculate the percentage of this energy that is converted to growth by the producers during photosynthesis. Show your working.

Answer is .... **(1 mark)**

4 Explain what happens to the energy reaching producers that is **not** converted to growth during photosynthesis.

> Will all the light fall on photosynthetic tissue?

> Sunlight is made up of many different wavelengths of light.

....

.... **(2 marks)**

5 Give four reasons why the rain forest has a higher GPP than the pine forest.

....

....

....

....

.... **(4 marks)**

Cycling of nutrients

1 Which of the groups below is made up of molecules which each contain carbon **and** nitrogen?

☐ **A** ornithine, glycogen, amylase, cholesterol, DNA

☐ **B** ornithine, collagen, albumen, NADP, cytosine

☐ **C** collagen, glycogen, ATP, triose phosphate, amino acids

☐ **D** collagen, albumen, insulin, amylose, phospholipid

> The ornithine cycle takes place in the liver and results in the production of urea.

(1 mark)

2 The table relates to the roles that microorganisms play in three processes involved in the nitrogen cycle. Use a cross (×) to indicate any statement that is incorrect.

Feature of cycle		Nitrogen fixation	Nitrification	Denitrification	Decomposition
The starting point is…	ammonium compounds				
	atmospheric nitrogen				
	nitrate				
	nitrogen-containing organic molecules				
The product is…	ammonia				
	gaseous nitrogen				
	nitrates				

(4 marks)

3 (a) Crop plants growing in areas of waterlogged soil in fields will often appear yellow and stunted compared with healthy looking plants in the rest of the field. Give two reasons why this might be so and explain your answer.

> Yellow stunted growth is a symptom of protein deficiency in plants.

....

....

....

....

.... **(3 marks)**

(b) Suggest why denitrifying bacteria are more active in waterlogged soil.

.... **(1 mark)**

4 Cellulose and starch are organic carbon-containing compounds. Describe the role of microorganisms in the recycling of the carbon from these compounds.

> For 'describe' questions, you do not need to make a judgement or explain how or why, you only need to describe what is requested

....

....

....

....

.... **(3 marks)**

Succession

1 Most biologists contend that abiotic factors are more important than biotic factors in determining which species are present during the colonisation stages of succession. Suggest why this is.

....

.... **(2 marks)**

2 Pioneer organisms are sometimes called 'opportunistic species' whereas those of late succession stages are described as 'equilibrium species'. Give two differences in the characteristics of these two groups of species.

> You could think about the relative sizes of individuals, the rate at which they reproduce and their ability to compete with other species.

....

.... **(2 marks)**

3 The diagram shows succession on the sand dunes at Gibraltar Point on the east coast of England.

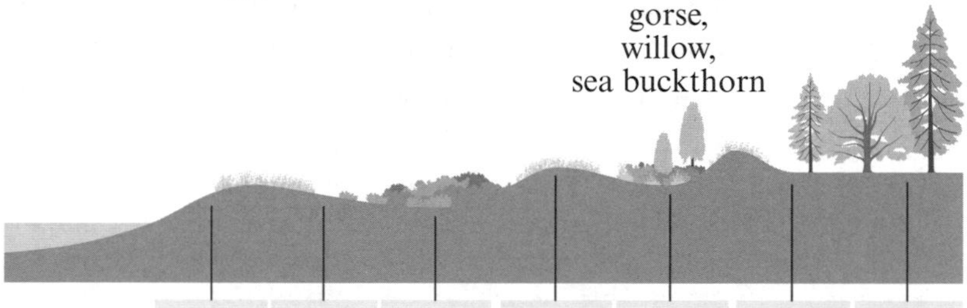

age (years)	0–65	65–95	95–125	125–185	185–245	245–365	>365
soil pH	6.6–7.0	4.8–5.5	3.9–4.9	3.9–4.5	3.9–4.5	3.6–4.5	4.5
soil colour		yellow		yellow-grey		grey	brown
humus (%)	0.2	0.6	2.5	3.2	8.2	13.5	>40

(a) Explain why the build-up of humus is important during succession.

> This is an 'explain' question, so it is not enough to state your point – you must also give reasons or say why you have come to that conclusion.

....

.... **(2 marks)**

Guided (b) Describe how primary productivity changes as succession proceeds.

Primary productivity depends on the amount of photosynthesis. As succession

proceeds, primary productivity ....

....

.... **(4 marks)**

Biotic and abiotic factors

1   A rock pool is a pool of sea water left
behind at low tide. They are found on
rocky shores.

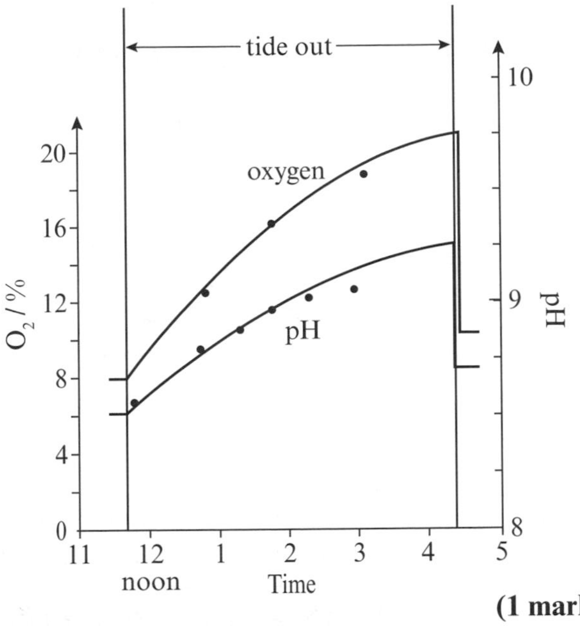

> Remember that seaweed (algae) is often
> abundant in a rock pool as are many
> invertebrates including sea anemones,
> snails, shrimps and crabs.

(a)   In the graph why do the oxygen levels
and the pH fall sharply to their original
values just after 4pm?

....

....

(1 mark)

(b)   If recordings had continued, how would you expect the oxygen and pH to have
changed over the next six hours?

.... **(1 mark)**

(c)   Suggest a reason for the rise in oxygen level during the afternoon.

.... **(1 mark)**

(d)   Suggest a reason for the rise in pH during the afternoon.

.... **(1 mark)**

(e)   Discuss the effects of the change in the two abiotic factors (O_2 and pH) on the
invertebrates in the pool.

> For 'discuss' questions, you should consider all aspects of the information,
> identify issues and explore them and give reasoned explanations.

....

....

.... **(2 marks)**

2   Sundew is a small insectivorous plant. It is adapted to grow in soils with a very low
pH and tends to be found in areas of bog where soils are waterlogged and cold.
Sundew is not found in areas where the soil pH is higher which is more typical of
many soils. Soils with low pH are very low in available nutrients such as nitrate and
phosphate ions. Discuss the possible reasons why sundew is only found in areas with
low pH, taking into account the information given here.

....

....

....

.... **(3 marks)**

143

Abiotic factors and morphology

Practical skills

Maths skills

1 Limpets (*Patella vulgata*) are a type of mollusc found on rocky shores. The shape of the shell differs in different areas of the shore and on shores with different amounts of exposure. Limpet shells may be relatively high domed or much flatter and broader. Scientists use the ratio of shell height to width to investigate this.

high domed limpet found on upper shore   broader flatter limpet found on lower shore

(a) Why is height to width ratio used rather than just the height or the width?

....

(1 mark)

Guided

(b) *A student noticed that the limpets on the upper and lower shore had different shapes. Suggest how this could be investigated.

> In this type of question, you need to consider the hypothesis, the independent variable, the dependent variable, the controlled variable(s), the uncontrolled variable(s) and the valid results.

Hypothesis: The shells of the limpets on the upper shore have a different height

to width ratio from those on the lower shore. ....

....

....

....

....

....

....

(6 marks)

(c) The results showed that the height to width ratios from the two sites overlapped, although the means were different. Suggest and explain two different statistical methods for demonstrating whether the difference between the two sites is significant.

....

....

....

(2 marks)

(d) Suggest a biological explanation for the observation that limpets higher up the shore have a significantly smaller height to width ratio.

> What is the biggest problem for marine organisms high up the beach?

....

(1 mark)

(e) A similar investigation was carried out to compare the shells of limpets from the middle shore zone of two rocky shores. One of these shores was more exposed to wave action than the other. In each case, the height to total size ratio was calculated. The limpets on the more sheltered shore were found to have a significantly higher height to width ratio. Suggest a reason for this.

....

(1 mark)

Human factors and climate change

1 Which of these techniques could be used to provide evidence of climate change?

   i   amniocentesis   iii peat bog pollen analysis
   ii dendrochronology   iv   potassium–argon dating

   ☐ **A** i and ii   ☐ **B** iii and iv   ☐ **C** i and iii   ☐ **D** ii and iii   **(1 mark)**

2 Which of the following are greenhouse gases?

   i   carbon dioxide   iii CFCs   v   nitrogen
   ii carbon monoxide   iv   methane   vi nitrogen oxides

   ☐ **A** i, iv, v and vi   ☐ **B** ii, iii, iv and vi   ☐ **C** i, iii, iv and v   ☐ **D** i, iii, iv and vi   **(1 mark)**

3 Many people have suggested that reducing the use of fossil fuels will help prevent further global warming. How does the greenhouse effect lead to higher global temperatures?

   > It is important to understand how this works so that you can then work out the effects of human activity.

   ....

   ....

   ....

   .... **(3 marks)**

4 Why is the greenhouse effect so important for organisms for life on Earth?

   .... **(1 mark)**

5 How does the fossil record provide evidence that the world has gone through an ice age?

   .... **(1 mark)**

6 *Many people have suggested that reducing the use of fossil fuels will help prevent further global warming. Discuss why some scientists may disagree with this suggestion.

   > This is a starred (*) question, so you need to structure your answer logically and make sure you use correct scientific terminology. There are many aspects to consider – start by addressing the ones directly related to the question.

   ....

   ....

   ....

   ....

   ....

   ....

   ....

   .... **(6 marks)**

Human factors and sustainability

1 Many species have become extinct over the centuries. Why should we worry about extinction in the 21st century?

....

.... **(1 mark)**

2 What is the difference between conservation and preservation?

....

.... **(1 mark)**

3 Why is it important that indigenous people have a say in any decisions made about conserving species in their area?

....

.... **(1 mark)**

4 What is CITES and why is it important?

....

....

.... **(2 marks)**

5 *Discuss the advantages and disadvantages of different approaches to animal conservation in areas where there has been significant habitat destruction.

In starred questions (*), structure your answer logically showing how the points you make are related to or follow on from each other. You need to select and apply relevant knowledge of biological facts or concepts to support the argument.

....

....

....

....

....

....

.... **(6 marks)**

Exam skills

1   The diagram shows the relationship between GPP (gross primary productivity), NPP (net primary productivity) and R (respiration) for an area of grassland. The efficiency of energy transfer for GPP to NPP for this grassland is 45%.

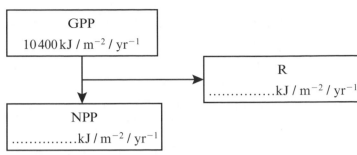

(a)   Explain the meaning of the term GPP.

....

....

.... **(2 marks)**

Maths skills

(b)   Calculate the values of NPP and R and add them to the diagram.

(2 marks)

(c)   Use the information given to explain the relationship between GPP and NPP.

....

....

....

.... **(3 marks)**

(d)   Suggest how a farmer who wanted to use this land for cattle might use this information.

> How will the number of cattle the farmer has relate to the NPP?

....

....

....

.... **(3 marks)**

(e)   The units used in the diagram (kJ / m^2 / yr) show a rate of energy production. Suggest why this is more useful than measurements of biomass in the grassland on a particular day.

....

....

.... **(2 marks)**

(f)   The farmer had another pasture with a different grass species. Here, the GPP was 10 800 kJ / m^2 / yr. Suggest reasons why the GPP is greater here than in the field mentioned above.

....

....

.... **(2 marks)**

Timed tests

Edexcel publishes official Sample Assessment Materials on its website. This timed test has been written to help you practise what you have learned and may not be representative of a real exam paper.

Remember that the official Edexcel specification and associated guidance materials are the authoritative source of information and should be referred to for definitive guidance.

AS Level timed test 1: Cellular Biology and Microbiology

1 hour and 30 minutes

Questions marked with an asterisk (*) will be marked on your ability to structure your answer logically, showing how the points you make are related to or follow on from each other.

1   Invertase is an enzyme used to create soft centres in chocolates. It does this by breaking down crystalline sucrose encased in chocolate over a period of a few days.

   (a) State the name of the reaction catalysed by invertase. **(1)**

   (b) State the names of all the reactants and products of this breakdown. **(2)**

   (c) Explain how the conditions might be controlled to make sure the breakdown of the sucrose takes a few days. **(2)**

   (Total for Question 1 = 5 marks)

2   The diagram shows the sequence of events leading to polypeptide synthesis.

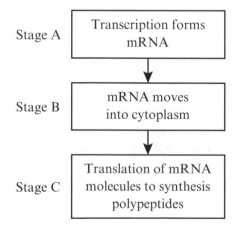

Stage A — Transcription forms mRNA

Stage B — mRNA moves into cytoplasm

Stage C — Translation of mRNA molecules to synthesis polypeptides

   (a) Tick the box next to the correct term that completes each of the following statements.

      (i) Translation takes place in or on the

         ☐ **A** Golgi apparatus   ☐ **C** nucleus

         ☐ **B** lysosome   ☐ **D** ribosome   **(1)**

   (ii) A triplet of bases that could not be found in mRNA is

     ☐ **A** adenine adenine guanine   ☐ **C** adenine cytosine guanine

     ☐ **B** adenine thymine guanine   ☐ **D** adenine uracil guanine   **(1)**

(b) Explain how proteins are made in cells.   **(4)**

(Total for Question 2 = 6 marks)

3   The diagram shows a β-glucose molecule which is the monomer of cellulose.

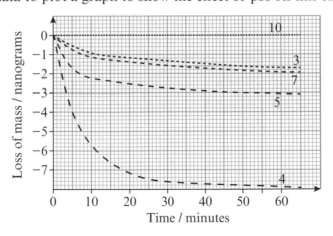

(a) Draw the products that are formed from a condensation reaction between two β-glucose molecules.   **(2)**

(b) Explain how the structure of cellulose makes it a suitable molecule to form the walls of cells.   **(3)**

(c) A study was carried out in which thin films of cellulose were put in contact with an enzyme, cellulase, which breaks cellulose down. The films were weighed at intervals and the loss in mass, due to cellulose digestion, was recorded. This was done at five pH values: 3, 4, 5, 7 and 10. The data are shown in the graph. Analyse the data to plot a graph to show the effect of pH on this enzyme.

    (4)

(Total for Question 3 = 9 marks)

4   A gene contained 450 bases of which 32% were guanine.

(a)   Calculate the number of uracil bases present in the mRNA made from this gene.   **(3)**

(b)   DNA is made up of the following components.

bases, e.g.
adenine   phosphate   ribose   deoxyribose

☐ A   ○   ⬠ R   ⬠ D

Draw the part of a DNA molecule which would give rise to the codon UAC in mRNA.   **(3)**

(c)   In their 1953 paper describing their model of DNA structure, Watson and Crick said:

> 'It has not escaped our notice that the specific pairing we have postulated immediately suggests a possible copying mechanism for the genetic material.'

Explain what they meant.   **(4)**

(Total for Question 4 = 10 marks)

5   Fertilisation happens in both mammals and flowering plants.

(a)   *Explain how, in mammals, events following the acrosome reaction prevent more than one sperm fertilising an egg.   **(6)**

(b)   A student investigated the effect of sucrose concentration on the growth of pollen tubes. Four pollen grains were placed in a small dish containing water. The pollen grains were left for two hours and the lengths of the pollen tubes produced were measured. The mean length was then calculated. This procedure was repeated using dishes containing sucrose solutions at concentrations of 5%, 10%, 20% and 30%. The student was told that his method was not valid and that some variables may not be controlled.

(i)   Describe three ways in which the student could modify the investigation to address these concerns.   **(3)**

(ii)   The graph shows the mean lengths of the pollen tubes. Error bars are also shown.

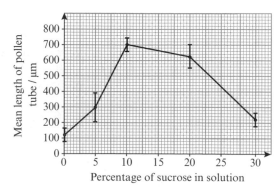

Analyse the data to describe the effect of increasing sucrose concentration on the mean length of pollen tubes over the two-hour period.   **(3)**

(Total for Question 5 = 12 marks)

6 (a) (i) LeJune, who stained chromosomes in human cells, first showed the cause
           of Down's syndrome in 1959. Name a stain that LeJune may have used.   **(1)**

       (ii) In which of the following stages of mitosis would chromosomes have been
            observed in this study?

            ☐ **A** interphase   ☐ **C** cytokinesis

            ☐ **B** metaphase   ☐ **D** telophase   **(1)**

       (iii) The diagram shows the karyotype of a normal person. Sketch in the
             change(s) you would expect in the karyotype of a boy with Down's
             syndrome.

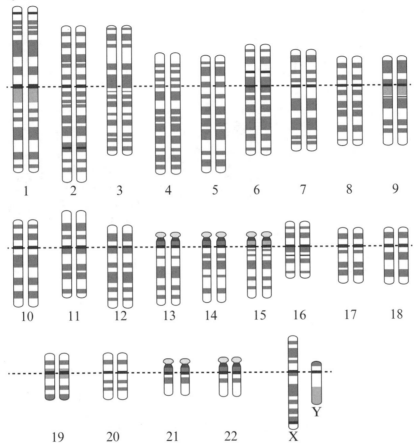

                                                                                    (1)

       (iv) Explain how Down's syndrome is caused.   **(3)**

    The table shows the relationship between the mother's age and the incidence of
    Down's syndrome.

Age of mother	Down's syndrome / 1000 births
25	0.8
30	1.0
35	3.0
40	10.0
45	20.0
50	80.0

    (b) Comment on the relationship between the age of the mother and the incidence
         of Down's syndrome based on the data given here.   **(4)**

                                                 (Total for Question 6 = 10 marks)

7 Carl Woese suggested that living organisms could be grouped into three domains: Bacteria, Archaea and Eukarya.

  (a) (i) Give one reason why Woese concluded that living organisms could be grouped into three domains. **(1)**

      (ii) When Carl Woese first suggested that all organisms could be classified into one of the three domains, his ideas were not accepted. Explain how Woese's idea was critically evaluated. **(3)**

  (b) (i) State why viruses are not classified in any of Woese's domains. **(1)**

      (ii) The diagram shows the sites of action of four different antiviral drugs, labelled A to D.

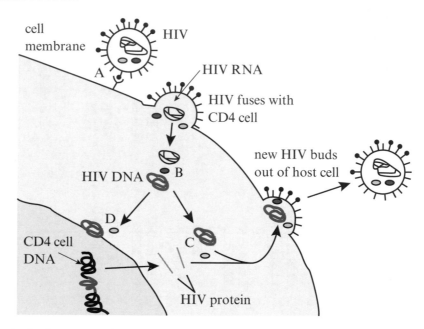

        Match the letters to the following drug types: reverse transcriptase inhibitors; attachment inhibitors; protease inhibitors; integrase inhibitors. **(2)**

     (iii) Explain why the Ebola virus has not yet caused a global epidemic. **(3)**

  (c) Compare and contrast the lytic and lysogenic cycles in viruses. **(4)**

(Total for Question 7 = 14 marks)

8 L-dopa forms a colourless solution in water. Dopa oxidase is an enzyme that converts L-dopa into dopachrome, which is red. A student used a colorimeter to investigate this reaction.

(a) (i) The graph shows the course of a reaction when the student used an enzyme concentration of 20 arbitrary units (AU).

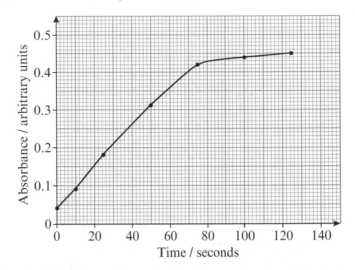

Calculate the initial rate of reaction for this concentration of enzyme. **(3)**

(ii) The student repeated the procedure using a range of enzyme concentrations. The results are shown in the table.

Enzyme concentration / AU	Initial rate of reaction / absorbance / 1000 s
0	0.0
10	2.5
30	6.1
50	9.0
70	11.0
90	11.0

Plot the data in the table on a graph. **(3)**

(iii) Give reasons for the effect of enzyme concentration on the initial rate of this reaction. **(3)**

(b) All enzymes are made from protein. In some cases, the protein molecule is modified and packaged before it is secreted. Describe how a protein is modified, packaged and secreted as an enzyme. **(5)**

(Total for Question 8 = 14 marks)

TOTAL FOR PAPER = 80 marks

AS Level timed test 2: Core Physiology and Ecology

1 hour and 30 minutes

Questions marked with an asterisk (*) will be marked on your ability to structure your answer logically, showing how the points you make are related to or follow on from each other.

1   The following questions are about transport in plants.

(a)   The pathway by which water moves across the root but does not enter living cells is called the

☐ **A** symplast   ☐ **B** tonoplast   ☐ **C** apoplast   ☐ **D** amyloplast   **(1)**

(b)   The cellular structure in the root, into which all water must pass, due to the Casparian strip, is called the

☐ **A** epidermis   ☐ **C** endodermis

☐ **B** pericycle   ☐ **D** cortex   **(1)**

(c)   Water moves into root hair cells by

☐ **A** osmosis   ☐ **C** facilitated diffusion

☐ **B** active transport   ☐ **D** endocytosis   **(1)**

(d)   A dendrometer, which can measure small changes in the diameter of woody structures such as twigs, was placed on a twig for six days. The results are shown in the graph.

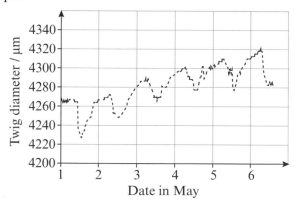

(i)   The overall rise in the curve is due to growth of the twig. Calculate the growth rate of the twig between 2 May and 6 May. Express your answer as mm / h and in standard form.   **(5)**

(ii)   Explain why the diameter of this twig does not increase smoothly.   **(4)**

(e)   Greenfly (aphids) feed on phloem sap and froghoppers feed on xylem sap. Both of these insects have a cibarial pump muscle in the head to assist in obtaining this sap. Deduce which of the two insects would have the largest, most powerful, cibarial pump muscle.   **(3)**

(Total for Question 1 = 15 marks)

2 (a) Photo A shows a child with a condition called kwashiorkor. His main food is likely to be fruit and rice. Photo B shows a person with a condition called filariasis elephantiasis, caused by a small worm, many of which live in the lymph capillaries.

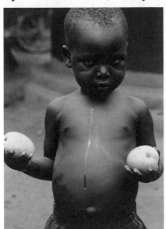

Photo A

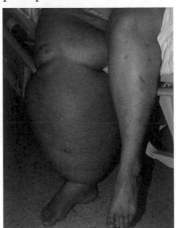

Photo B

Compare and contrast the cause of each of these conditions.   (4)

(b) Blood clots can form if the lining of an artery becomes damaged. One cause of a stroke is a blood clot in an artery that supplies the brain with blood.

(i) State the main stages of the blood-clotting process.   (5)

(ii) Explain the changes within the artery wall that can lead to a person suffering a stroke.   (4)

(Total for Question 2 = 13 marks)

3 Osteocalcin is a structural protein found in the bones of mammals. The sequence of the amino acids in osteocalcin was determined. The first 20 amino acids of osteocalcin from humans and some apes are shown in the table. Each is represented by a capital letter.

Mammal	Amino acid number																			
	1				5					10					15					20
human	Y	L	Y	Q	W	L	G	A	P	V	P	Y	P	D	P	L	E	P	R	R
chimpanzee	Y	L	Y	Q	W	L	G	A	P	V	P	Y	P	D	P	L	E	P	R	R
orang-utan	Y	L	Y	Q	W	L	G	A	P	V	P	Y	P	D	P	L	E	P	K	R
gorilla	Y	L	Y	Q	W	L	G	A	O	V	P	Y	P	D	P	L	E	P	K	R

(a) Analyse the data to explain the evolutionary relationships between humans and these apes.   (4)

(b) An analysis of the DNA of each species could be used as another source of data about the evolutionary relationships between them.

(i) Describe how genetic fingerprinting could be carried out on a sample from any one of the species shown.   (5)

(ii) Explain how such an analysis could provide further evidence to support the conclusions from the amino acid sequences.   (2)

(Total for Question 3 = 11 marks)

4 (a) A 10-year survey was carried out on the species richness of hedgerows and roadside verges. The data are shown in the bar charts.

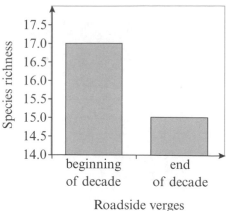

Roadside verges

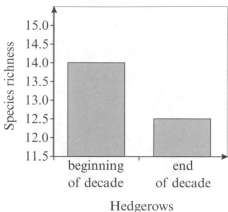

Hedgerows

   (i) Analyse the data to compare the changes in these two habitats over the 10-year period. **(4)**

   (ii) Criticise this study, which was entitled 'Biodiversity changes in hedgerows and roadside verges over a 10-year period'. **(4)**

(b) Discuss the pros and cons of seed banks in the conservation of plant species and reasons for continuing to fund them. **(5)**

(Total for Question 4 = 13 marks)

5 The photograph shows a waxy leaf frog (*Phyllomedusa sauvagii*). This species of frog is found in hot, dry areas of South America. It has glands that produce waxy lipids to spread over its skin. This reduces water loss. The waxy leaf frog is active only at night, when it hunts for insects in the trees.

(a) (i) State how the waxy leaf frog is physiologically adapted to its environment. **(1)**

   (ii) Describe a behavioural adaptation of the waxy leaf frog to its environment. **(1)**

   (iii) Give reasons why the behavioural adaptation described enables the waxy leaf frog to survive in this habitat. **(1)**

(b) Describe the niche of the waxy leaf frog. **(2)**

(c) Discuss how natural selection could have given rise to the adaptations shown by the waxy leaf frog. **(5)**

(Total for Question 5 = 10 marks)

6   Fenitrothion is an insecticide used to control insects that feed on crop plants. Before an insecticide is approved for use, its effects on insects and other animals are tested. The testing of fenitrothion showed that it affects the permeability of animal cell membranes. Some scientists investigated the effect of fenitrothion on the permeability of the plant cell membranes of beetroot. The diagram shows a beetroot cell with a vacuole containing a red pigment called betalain.

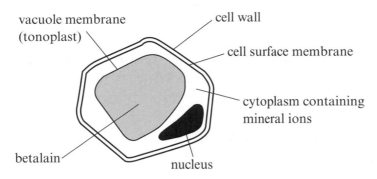

Discs were cut from a beetroot. Any betalain on the outside of the discs was removed by washing the discs in water. Twenty discs were placed into a beaker containing 20 cm³ of fenitrothion solution in water. Betalain began to leak from the discs, changing the colour of the solution. The colour of the solution in the beaker was recorded every hour. In another experiment, the scientists investigated the movement of minerals out of the cells over the same time period. The graph shows their results.

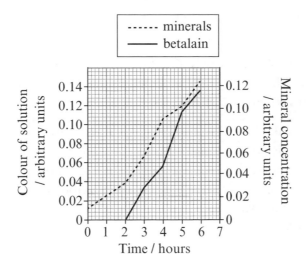

(a) Compare and contrast the patterns of movement of minerals and betalain pigment based on your analysis of the data provided.   **(4)**

(b) *When designing the experiment, the variables which need to be controlled were considered. Scientists thought pH might affect permeability. Devise a way to modify this experiment to determine the pH value that would have the least effect on the permeability of the membranes.   **(6)**

(Total for Question 6 = 10 marks)

7   (a) *'The heart is structurally well adapted to its function in the cardiovascular system.' Justify this statement using three examples.   **(6)**

(b) Describe how the complete cardiac cycle is controlled.   **(4)**

(Total for Question 7 = 10 marks)

TOTAL FOR PAPER = 82 marks

A Level timed test 1: Advanced Biochemistry, Microbiology and Genetics

1 hour 45 minutes

Questions marked with an asterisk (*) will be marked on your ability to structure your answer logically, showing how the points you make are related to or follow on from each other.

1   The diagram shows a β-glucose molecule which is the monomer of cellulose.

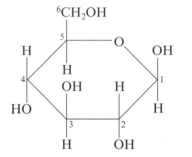

(a) Draw the products that are formed from a condensation reaction between two β-glucose molecules. **(2)**

(b) Name the bond formed in the condensation reaction between two β-glucose molecules. **(1)**

(c) Explain how the structure of cellulose makes it a suitable molecule to form the walls of cells. **(4)**

(d) Prokaryotic and eukaryotic cells differ in more than just the composition of their cell walls. Which of A, B, C or D shows the correct ultrastructure of prokaryotic and eukaryotic cells?

	Prokaryotic	Eukaryotic
☐ A	peptidoglycan cell wall mitochondria 80S ribosomes	peptidoglycan cell wall mitochondria 70S ribosomes
☐ B	cellulose cell wall centrioles 70S ribosomes	cellulose cell wall mitochondria 70S ribosomes
☐ C	peptidoglycan cell wall nucleoid 80S ribosomes	cellulose cell wall nucleolus 80S ribosomes
☐ D	peptidoglycan cell wall nucleoid 70S ribosomes	cellulose cell wall nucleolus 80S ribosomes

(1)

(Total for Question 1 = 8 marks)

2   The Human Genome Project was started in 1990 and completed in 2003. It aimed to sequence the human genome.

(a) State what is meant by the term genome. **(1)**

(b) DNA can be amplified using the technique called PCR. DNA can be replicated inside a nucleus. State two similarities and two differences between these processes. **(4)**

(Total for Question 2 = 5 marks)

3 People can receive vaccinations against diseases which cause levels of certain antibodies to increase in their blood. During the 1970s there was an intensive smallpox vaccination programme.

   (a) Name the immune cells that produce antibodies.

     ☐ **A** B effector cells   ☐ **C** plasma cells

     ☐ **B** macrophages   ☐ **D** T helper cells   **(1)**

   (b) *Explain how a smallpox vaccination could lead to immunity.   **(6)**

   (c) The graph shows the concentration of smallpox antibodies in the blood following a vaccination.

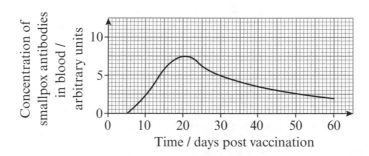

   (i) State the type of immunity gained from this vaccination.

     ☐ **A** active artificial   ☐ **C** passive artificial

     ☐ **B** active natural   ☐ **D** passive natural   **(1)**

   (ii) Draw a new line on the graph to show the concentration of antibodies in the blood if the vaccinated individual was infected with the smallpox pathogen a few months later.   **(3)**

   (iii) Explain why there is a difference between the levels of antibodies produced after vaccination and the levels of antibodies when the vaccinated individual was then infected with the smallpox pathogen.   **(4)**

(Total for Question 3 = 15 marks)

4 (a) *E. coli* bacteria can be used in biotechnology to produce clotting factors for haemophiliacs. The gene coding for factor VIII can be transferred into the bacterial DNA.

   (i) Describe how the gene is inserted into the bacterial DNA.   **(3)**

   (ii) The DNA formed is called recombinant DNA. State a method that could be used to insert the recombinant DNA into the bacterial cell.   **(1)**

   (b) A biotechnology company would want to check to see if all the *E. coli* bacteria contained the recombinant DNA before they placed the bacterial cells in a fermenter. Describe how the company could identify recombinant *E. coli* cells.   **(4)**

(c) The recombinant *E. coli* were placed in a fermenter for 60 hours. The graph shows the number of bacteria cells present during this time.

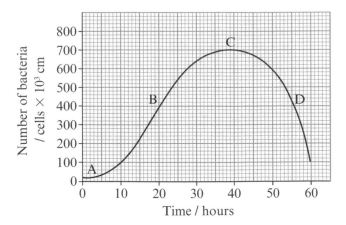

(i) Which letter identifies the log phase?

☐ A ☐ B ☐ C ☐ D **(1)**

(ii) Identify the length of the lag phase. **(1)**

(iii) Calculate the gradient at point B on the graph. **(2)**

(iv) Give a reason why this gradient is different from the gradient at point D. **(1)**

(Total for Question 4 = 13 marks)

5 Plant cells can divide to form identical cells in a process called mitosis. The diagram shows garlic root cells as viewed under a light microscope.

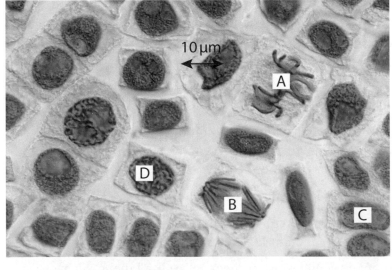

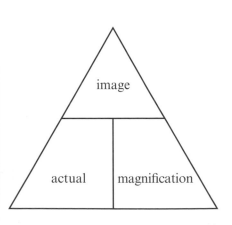

(a) Calculate the magnification of the cell labelled B. **(2)**

(b) (i) Which letter shows a cell that is in metaphase?

☐ A ☐ B ☐ C ☐ D **(1)**

(ii) Which letter shows a cell that is in telophase?

☐ A ☐ B ☐ C ☐ D **(1)**

(c) Two hundred cells from two different parts of a garlic root were analysed under a microscope to see if they were in interphase or undergoing mitosis. The results are shown in the table.

Stage of cell cycle	Number of cells in each stage of mitosis	
	Slide A	Slide B
interphase	155	200
prophase	21	0
metaphase	11	0
anaphase	6	0
telophase	7	0

  (i) Calculate the percentage of cells that were in metaphase in slide A. **(2)**

  (ii) Identify which slide was made from a section of root furthest from the tip. Justify your answer. **(3)**

(d) Gametes are produced by the process of meiosis. Explain why the DNA of one man's sperm cell is different from his other sperm cells and body cells. **(3)**

(Total for Question 5 = 12 marks)

6 (a) The diagram shows a cross section through the stem of a dicotyledon plant.

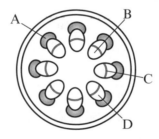

  (i) Where would lignin be found?

   ☐ A   ☐ B   ☐ C   ☐ D   **(1)**

  (ii) Where would companion cells be found?

   ☐ A   ☐ B   ☐ C   ☐ D   **(1)**

(b) Glucose can be converted by plant cells into sucrose for transport in phloem tissues. Explain the role of the companion cells in the movement of sucrose molecules into phloem. **(3)**

(c) Garlic and onions can convert glucose into fructose. The fructose molecules are stored as polymers called fructans. Explain why the plant cells need to convert fructose to fructans for storage inside a cell. **(3)**

(d) *A student thought that the water potential of cells in a seed would change as the seed started to germinate because the starch was being hydrolysed into glucose. She wrote the following hypothesis: 'Germinating seeds will have a lower water potential than non-germinating seeds.' Devise an investigation to test this hypothesis and collect valid data. **(6)**

(Total for Question 6 = 14 marks)

7 (a) Haemoglobin is a globular protein that can be found in red blood cells. The enzyme carbonic anhydrase is also a globular protein found in red blood cells.

    (i) Name the bond which is formed between two amino acids when they are joined together in a condensation reaction.

      ☐ **A** ester

      ☐ **B** glycosidic

      ☐ **C** hydrogen

      ☐ **D** peptide   **(1)**

    (ii) Compare and contrast the structure of haemoglobin and enzymes such as carbonic anhydrase.   **(5)**

(b) The graph shows the percentage saturation of haemoglobin at different partial pressures of oxygen. The percentage saturation of haemoglobin can be affected by many factors and can give rise to the curves labelled A and B on the graph.

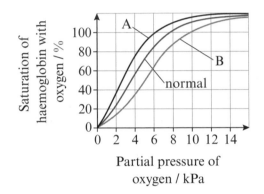

    (i) How many molecules of oxygen can one molecule of haemoglobin carry when it is fully saturated?   **(1)**

    (ii) Which curve, A or B, would be produced if the haemoglobin was from a fetus? Justify your answer.   **(3)**

(c) The enzyme carbonic anhydrase catalyses a reaction between carbon dioxide and water in red blood cells. A non-competitive inhibitor drug called acetazolamide can inhibit this reaction. Complete the table to compare competitive and non-competitive inhibition.

	Competitive	Non-competitive
Binding site		
Effect on active site		
Effect on rate of reaction		

(3)

(Total for Question 7 = 13 marks)

8 The graph shows the changes in blood pressure in the left hand side of the heart during the cardiac cycle.

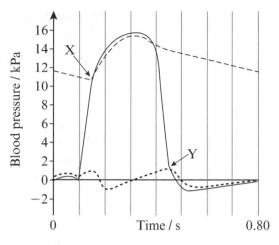

(a) Calculate the heart rate in beats per minute. (2)

(b) Which row, A, B, C or D, correctly shows what is happening to the valves in the heart at points X and Y on the graph?

	X	Y
☐ A	atrioventricular valve closed semilunar valve open	atrioventricular valve closed semilunar valve open
☐ B	atrioventricular valve open semilunar valve closed	atrioventricular valve open semilunar valve closed
☐ C	atrioventricular valve closed semilunar valve open	atrioventricular valve open semilunar valve closed
☐ D	atrioventricular valve open semilunar valve closed	atrioventricular valve closed semilunar valve open

(1)

(c) Describe the events in the heart that have led to the increase in ventricular pressure between 0.10 and 0.15 seconds. (3)

(d) The pressure in veins is always lower than the pressure in the aorta. Explain how the structure of veins is related to their function. (4)

(Total for Question 8 = 10 marks)

TOTAL FOR PAPER = 90 marks

A Level timed test 2: Advanced Physiology, Evolution and Ecology

> **1 hour 45 minutes**
>
> Questions marked with an asterisk (*) will be marked on your ability to structure your answer logically, showing how the points you make are related to or follow on from each other.

1   The scientist Carl Woese published an article in a scientific journal in 1977 suggesting that living organisms could be grouped into three domains. There are specific differences between the organisms in the three domains.

   (a) (i)   Describe how the scientific community would have evaluated Woese's theory. **(2)**

      (ii)   Which statement correctly identifies the names of the three domains suggested by Woese.

         ☐ **A** Animalia, Archaea and Eukarya

         ☐ **B** Animalia, Bacteria and Prokaryotae

         ☐ **C** Archaea, Bacteria and Eukarya

         ☐ **D** Archaea, Eukarya and Prokaryotae **(1)**

   (b)   The diagram shows a typical animal cell. Which organelle would you expect to find more of in a steroid hormone releasing cell?

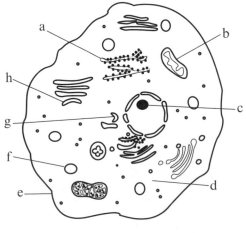

         ☐ **A** a   ☐ **B** c   ☐ **C** g   ☐ **D** h **(1)**

   (c)   In which two parts of the cell could you expect to find enzymes?

         ☐ **A** b and d   ☐ **C** d and e

         ☐ **B** b and g   ☐ **D** d and g **(1)**

(d) The cytoplasm of a cell contains a lot of water. Water is described as a dipolar molecule because it has a

   ☐ **A** positively charged hydrogen end and a negatively charged oxygen end

   ☐ **B** positively charged hydrogen end and a positively charged oxygen end

   ☐ **C** negatively charged hydrogen end and a negatively charged oxygen end

   ☐ **D** negatively charged hydrogen end and a positively charged oxygen end **(1)**

(e) An experiment was carried out to compare the uptake of substance A and B into prokaryote cells. Some cells were placed in a solution containing equal concentrations of both substances and kept at 25 °C. The cytoplasmic concentration of both substances was measured over five hours and are shown in the graph.

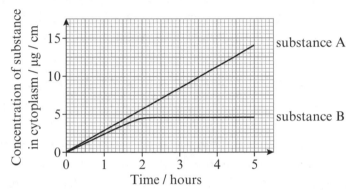

(i) Calculate the percentage increase of substance B in the cytoplasm of the cell between one and two hours. **(2)**

(ii) Substance B enters the cells by diffusion. Explain how the results of this experiment support this statement. **(3)**

(iii) Substance A enters the cells by active transport. Give two differences between active transport and diffusion. **(2)**

(Total for Question 1 = 13 marks)

2 The retina contains rod and cone cells.

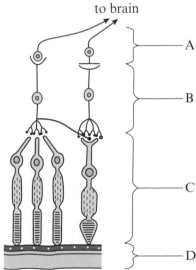

(a) (i) When light stimulates a rod cell the pigment changes. Name this pigment.

   ☐ **A** iodopsin   ☐ **C** phytochrome red

   ☐ **B** phytochrome far red   ☐ **D** rhodopsin **(1)**

(ii) After changing, the pigment takes time to become functional again. This is because

☐ **A** it has to bleach

☐ **B** the membrane has to be polarised

☐ **C** the rod cell needs to reset

☐ **D** two components have to be rejoined   **(1)**

(iii) The cell that links a rod cell to a sensory neurone is

☐ **A** a bipolar neurone   ☐ **C** a unipolar neurone

☐ **B** a multipolar neurone   ☐ **D** an optic nerve   **(1)**

(b) When light reaches a rod cell the voltage across the cell surface membrane can change. This can lead to the formation of an action potential in an optic neurone. Explain how light causes a change in the voltage across the cell surface membrane of a rod cell.   **(4)**

(c) Explain why cone cells are less sensitive to light than rod cells.   **(2)**

(Total for Question 2 = 9 marks)

3   Light energy is absorbed by a producer, transformed into chemical energy and some is used to form biomass. Consumption of this biomass would pass the energy from one trophic level to another. Some energy is not passed from one trophic level to another in an ecosystem.

(a) State what is meant by the term trophic level.   **(1)**

(b) Using the diagram, calculate the percentage of energy in trophic level 1 transferred to new biomass in trophic level 2.

Trophic level 1
energy in biomass = 5300 kJ

⇨

Trophic level 2
energy of food ingested = 2800 kJ
energy lost in respiration, urine and faeces = 1750 kJ

(2)

(c) The table shows some terms used to describe organisms that can be found in ecosystems. Complete the table by adding ticks (✓) to show which trophic level(s) match each term.

	Trophic level 1	Trophic level 2	Trophic level 3	Trophic level 4
Autotroph				
Carnivore				
Herbivore				
Heterotroph				
Primary consumer				
Tertiary consumer				

(3)

(d) The graphs show the relationship between net primary production and two abiotic factors in a habitat.

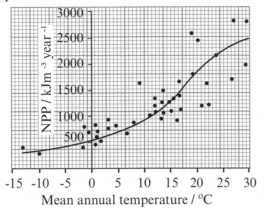
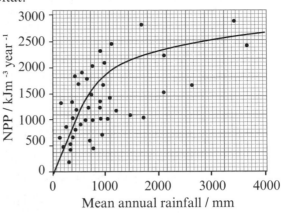

(i) Explain the meaning of the term net primary productivity (NPP). **(2)**

(ii) Analyse the information in the graphs to explain the relationship between NPP and each of these two environmental factors. **(4)**

(Total for Question 3 = 12 marks)

4 Nerve impulses are transmitted along the axon of a neurone.

(a) The diagram shows the structure of a motor neurone.

(i) Name the part of the neurone labelled T.

☐ **A** dendrite

☐ **B** node of Ranvier

☐ **C** Schwann cell

☐ **D** synapse **(1)**

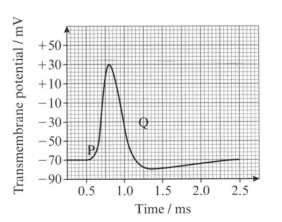

(ii) The graph shows changes in the membrane potential during the transmission of an impulse along the axon of a motor neurone. Choose the description of the membrane potential at 0.75 ms on the graph.

☐ **A** depolarised

☐ **B** hyperpolarised

☐ **C** polarised

☐ **D** repolarised **(1)**

(b) Complete the table below by inserting a tick to show which ions are able to move across the membrane at points P and Q on the diagram. **(2)**

Point on diagram	Permeable to potassium ions	Permeable to sodium ions
P		
Q		

(c) The cells of the Calabar bean (*Physostigma venenosum*) contain a chemical called physostigmine. This chemical inhibits the enzyme acetylcholinesterase.

(i) State the genus to which the Calabar bean belongs. **(1)**

(ii) The physostigmine chemical derived from the bean cells inhibits the enzyme acetylcholinesterase. Explain how this would affect the transmission of a nerve impulse across a synapse. **(6)**

(Total for Question 4 = 11 marks)

5 Analysis of several regions of the DNA of 31 different cheetahs showed very little genetic variation.

  (a) Explain how the DNA from different cheetahs can be analysed to show the presence or absence of genetic variation. **(4)**

  (b) Explain why a low genetic diversity could be a problem if the environment changes. **(2)**

  (c) A female cheetah often mates with several different males and gives birth to two or three cubs at a time, each having a different father. Give reasons why this may be advantageous to cheetahs. **(2)**

  (d) Give two differences between genetic diversity and species richness. **(2)**

The Hardy–Weinberg equation states that allele and genotype frequencies in a population will remain constant from one generation to the next if there are no other influencing factors. It is calculated using the following equation:

$$p^2 + 2pq + q^2 = 1$$
$$p + q = 1$$

  (e) Cheetahs only have 4% of genes that have more than one allele. 98 out of 200 cheetahs have the homozygous recessive genotype for a particular gene which has two alleles. Predict what percentage of the population would be heterozygotes. **(3)**

(Total for Question 5 = 13 marks)

6 (a) Oestrogen is a hormone that affects transcription. It forms a complex with a receptor in the cytoplasm of target cells. Explain how an activated oestrogen receptor affects the target cell. **(3)**

  (b) Oestrogen only affects target cells. Explain why oestrogen does not affect other cells in the body. **(1)**

  (c) ADH is another hormone that is secreted by the body. It is secreted by the

    ☐ **A** hypothalamus   ☐ **C** pancreas

    ☐ **B** kidney   ☐ **D** pituitary **(1)**

  (d) A kidney filtered 180 litres of filtrate a day. In the absence of antidiuretic hormone (ADH), 13% of the filtrate entered the collecting duct. In the presence of ADH, only 5.5% of the filtrate sometimes enters the collecting duct.

   (i) Calculate how much more of the filtrate is reabsorbed back into the bloodstream in the presence of ADH. **(2)**

   (ii) Explain how the presence of ADH results in more filtrate being reabsorbed back into the bloodstream. **(3)**

(Total for Question 6 = 10 marks)

7   The mean density of two plant species (A and B) was measured at different distances from a main road. The results of the investigations are shown in the kite diagram.

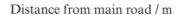

Distance from main road / m

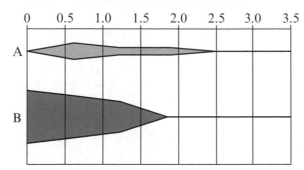

(a)   Describe a procedure that the student could have used to determine the mean density of plant species A and B.   **(3)**

(b)   Describe the relationship between the mean density of plants and distance from the road for species A and B.

Species B is called *Cochlearia danica* and its normal habitat is cliff-tops, sand dunes and sea-walls. Scientists who carried out the survey wrote a hypothesis: '*Cochlearia danica* is able to survive and grow near to main roads because it is tolerant to the high salt concentrations resulting from gritting.'   **(3)**

(c)   *Design a laboratory experiment to test this hypothesis.   **(6)**

(Total for Question 7 = 12 marks)

TOTAL FOR PAPER = 80 marks

ANSWERS

The answers given here are examples of possible responses. In some cases other answers may also be possible.

*6- and *9-mark questions

Throughout this book you will find 6-mark and 9-mark questions. These questions are starred (*). You should organise your ideas for these questions in a logical order.

You need to apply your knowledge and understanding of material to the qualities and skills outlined in the relevant mark scheme. You should also reach a decision/judgement on whatever the question asks. Below is a generic mark scheme for the 6- and 9-mark questions in this book.

	6-mark questions	9-mark questions	Descriptor
0	0	0	No rewardable material
1	1–2	1–3	• Demonstrates isolated elements of biological knowledge and understanding, with generalised comments • Limited analysis, interpretation and/or evaluation of the scientific information • Provides little or no reference to a range of scientific ideas, processes, techniques and procedures • Scientific argument may be attempted, but fails to link biological concepts and/or ideas in order to support decision/conclusion/judgement with evidence • Basic information with some attempt to link knowledge and understanding to the context
2	3–4	4–6	• Demonstrates adequate knowledge and understanding with selection of some biological facts/concepts to support the argument or decision/conclusion being made • Occasional evidence of analysis, interpretation and/or evaluation • Some linkages and lines of scientific reasoning • Scientific reasoning occasionally supported through the linkage of a range of scientific ideas, processes, techniques and procedures • Scientific argument/judgement is partially developed through the application of relevant evidence • Attempts to synthesise and integrate relevant knowledge with linkages to biological concepts and/or ideas, leading to a notional scientific argument or decision/conclusion based on evidence • Some structure
3	5–6	7–9	• Demonstrates comprehensive knowledge and understanding by selecting and applying relevant knowledge of biological facts/concepts to support the argument or decision/conclusion being made • Answer supported throughout by evidence from the analysis, interpretation and/or evaluation of the scientific information • Line(s) of argument/scientific reasoning supported throughout by sustained application/linkage of relevant evidence (scientific ideas, processes, techniques and procedures) • Well-developed, logical and sustained line of reasoning • Demonstrates throughout the skills of synthesising and integrating relevant knowledge with consistent linkages to biological concepts and/or ideas, leading to nuanced and balanced scientific argument or decision/conclusion based on evidence • A clear, coherent and logical structure

1. Carbohydrates 1

1 A (1)
2 two glucose molecules correctly drawn;
   indication that water is formed;
   glycosidic bond correctly drawn (3)
3 (3)

Disaccharide	Component monosaccharides	Type of bond between monosaccharides
sucrose	glucose and fructose	α-1,2 glycosidic
lactose	glucose and galactose	β-1,4 glycosidic
maltose	2 α-glucose	α-1,4 glycosidic

4 both hexose, both six-membered rings
   α has −OH below the ring at carbon atom 1 (of drawing);
   in β-glucose it is above the ring (2)

2. Carbohydrates 2

1 Both glycogen and starch are made from glucose molecules which are joined by 1,4 glycosidic bonds. In addition, 1,6 glycosidic bonds are found in amylopectin and glycogen. These bonds are easily broken by hydrolysis. All the molecules coil and thus occupy less space than the same number of glucose molecules would. They are insoluble and have no osmotic effect. (4)

2 (5)

Statement	Correct
polymer of glucose	✓
molecule contains α and β-glucose	
glycosidic bonds present	✓
molecule may have side branches	
molecule can form hydrogen bonds with adjacent molecules	✓

3 Amylose is unbranched (because it has no 1,6 glycosidic bonds) so there are fewer places for it to be broken down – just either end of the molecule. Amylopectin has many branches (due to 1,6 glycosidic bonds) and so there are more places where the glucose can be broken off, leading to a faster release of glucose and, therefore, energy. (2)

4 chains of β-glucose molecules held together by hydrogen bonds (3)

3. Lipids

1 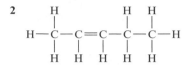 **(2)**

2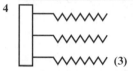

one mark for showing a double bond anywhere along the chain;
one mark for showing the **two** missing hydrogens **(2)**

3 **(1)**

Statement	Correct
Triglycerides are the building blocks of proteins.	
Triglycerides can be modified to form phospholipids.	✓
Water is released when a triglyceride molecule is hydrolysed.	
Some triglycerides contain sulfur.	

4 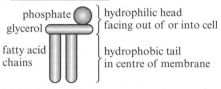 **(3)**

4. Functions of lipids

1 outside or inside cell

phosphate
glycerol } hydrophilic head facing out of or into cell

fatty acid chains } hydrophobic tail in centre of membrane **(5)**

2 Lipids are good for energy storage because they yield 2× as much energy as both carbohydrates and proteins. This is because energy is released when bonds form and more C–O bonds are made when a lipid is oxidised than when a carbohydrate or protein is. Lipids are insoluble, which stops them being washed out of living things and lost. They are good insulators because water-charged ions cannot get through a layer of them and this stops the flow of charge. **(4)**

3 The heads of phospholipids are hydrophilic and are attracted to water. The fatty acid tails are hydrophobic and orientate themselves away from water. **(3)**

4 Add more cholesterol and more unsaturated fat to the lipid bilayer of the membrane. **(2)**

5. Proteins: amino acids and polypeptides

1 **B (1)**

2 (a)

(b) dipeptide **(1)**
(c) They are not the same because they have different R groups, the R group differentiates one amino acid from another. **(2)**

3 **C (1)**

4 **A (1)**

6. Protein structures

1 *Primary structure is the sequence of amino acids in the chain. For example, in this case, Lys-Glu-Thr-Ala-Ala-Lys etc.
Secondary structure is the part of the structure stabilised by H bonds between the backbone of the molecule. In this case, there would be a region around D.

Tertiary structure is the structure which is stabilised by interactions between R side chains. Here, the Cys-Cys held together by disulfide bridges would be an example. **(6)**

2 A is carboxyl, C is amino **(2)**

7. Haemoglobin and collagen

1 Fibrous proteins are straight chains whereas globular are spherical.
Globular proteins are folded whereas fibrous proteins are not. Finally, globular proteins are soluble in water and but fibrous proteins are not. **(3)**

2 Collagen is made up of amino acids in a repeating sequence of three, one of which is always glycine. The other two are commonly proline and hydroxyproline. This primary structure is twisted into a helix, held together by H bonds between parts of the molecule's backbone. Three such helices are twisted together to form a triple superhelix held together by H bonds between the R groups of the amino acids. **(5)**

3 Haemoglobin is globular, so it is soluble in water.
It has four haem groups, which allows it to bind to four molecules of oxygen.
It undergoes a change in shape when it binds with a molecule of oxygen and the reverse change when it loses it. The shape change makes it have a greater affinity for oxygen as it gains it and a lower affinity as it loses it. **(3)**

4 There are four polypeptides in the quaternary structure. Each one can bind to oxygen.
When it does, the whole shape of the molecule changes and makes bonding of the next oxygen easier. **(3)**

8. Nucleic acids: DNA

1 **A (1)**

2 phosphate

pentose sugar

organic nitrogenous base **(3)**

3 The phosphate group is attached to the deoxyribose sugar at carbon atom number five by a condensation reaction and a base is attached to the deoxyribose sugar at carbon atom number one by a condensation reaction. **(3)**

4 A pairs with T; G pairs with C **(2)**

5 A purine with a purine would be too small and a pyrimidine with a pyrimidine would be too big, the bonding opportunities are incorrect. **(2)**

9. Nucleic acids: RNA

1 H bonds form in both between bases pairs. In DNA, they hold two separate strands in a double helix. In tRNA, they stabilise the folding of a single strand into a complex shape, often described as a clover leaf. In DNA, every base is H bonded to another base. In tRNA, many bases are not H bonded. **(3)**

2 **A (1)**

3 **B (1)**

4 **C (1)**

5 (a) tRNA **(1)**
(b) A is amino acid binding site
B is hydrogen bonds
C is anticodon **(3)**

10. DNA replication and the genetic code

1 Both a gene and a codon are a sequence of bases. However, a gene is many bases and codes for a polypeptide chain and a codon is a sequence of three bases that code for an amino acid. **(3)**

2 Degeneracy is where amino acids are coded for by more than one triplet codon. This means that some mutations may have no effect on the amino acids produced. **(2)**

3 Step 1 is C; Step 2 is D; Step 3 is A; Step 4 is B **(2)**

4 **C (1)**

11. Protein synthesis

1 (a) X is DNA; Y is mRNA; Z is protein **(3)**

(b) Transcription is when one of the strands of DNA is used as a template for the synthesis of mRNA. Ribonucleotides are paired with their complements on the template strand. Uracil instead of thymine pairs with adenine. RNA polymerase then joins up the ribonucleotides.

Translation is when a protein is synthesised by tRNA's bases pairing with the complementary codon on the mRNA, bringing designated amino acids with them. These amino acids then bond to each other to create a polypeptide chain. **(5)**

2 A polypeptide chain will need to be folded to create a globular shape which is precise for its function as an enzyme. **(2)**

12. Mutation

1 A gene is a sequence of bases that codes for a sequence of amino acids in a protein.

A mutation is a change in that sequence, so this may result in a different amino acid being inserted into the polypeptide chain

This may change the shape of the protein. Since enzyme function depends on protein shape, this differently shaped protein would not function as an enzyme. **(4)**

2 (a) If the mutation caused the mutated codon to become a stop codon rather than coding for an amino acid, then protein synthesis would stop at this point. **(2)**

(b) If the mutation caused the mutated codon to code for a different amino acid, a protein would be formed but it would fold imperfectly and thus not do its intended job. **(2)**

13. Sickle cell anaemia

1 **D (1)**

2 The condition sickle cell anaemia is caused by a mutation in the gene for the beta globin portion of the haemoglobin molecule. This mutation leads to the substitution of the amino acid valine for the amino acid glutamic acid. Glutamic acid is ionisable and soluble whereas valine is not. The haemoglobin formed using valine assumes a shape which causes individual molecules to join together in chains inside the red blood cells. This in turn causes their shape to change. **(4)**

3 Symptom 1: anaemia
Explanation: The damage to the red blood cells will result in less oxygen available and so less energy available.
Symptom 2: frequent infections
Explanation: damage to the spleen by sickle cells **(4)**

14. Enzymes

1 **C (1)**

2 (a) One of the features of enzymes is that they are specific to their substrate. Caffeine is the substrate in this case. Because the products are different shapes, P450 must consist of at least three enzymes with different active sites that will join with caffeine but which then lead to a different reaction and a different product. **(3)**

(b) $84 \div 12 = 7$
so $(50 \div 7 = 7.14\,mg)$ of theobromine would be produced
4% theophylline would be $\frac{1}{3}$ of this, $7.14 \div 3 = 2.38\,mg$ **(3)**

15. Activation energy and catalysts

1 **B (1)**

2 Enzymes can put a strain on bonds in the substrate to break them. They may also provide a more favourable pH in the active site. Finally, they can bring reactants close together in the active site so bonds are easier to form. **(3)**

3 A is the energy rise in the substrate needed to get the reaction to happen when an enzyme is present. It is much lower than B.
B is the energy rise in the substrate needed to get the reaction to happen without an enzyme, e.g. by heating it.

C is the difference in the energy possessed by the substrate minus the energy possessed by the products, this is given out as heat. **(3)**

16. Reaction rates

1 Enzyme reactions are very fast and the substrate, in this case hydrogen peroxide, is rapidly used up and so the rate changes. This means that measuring something to an end-point, as is suggested in this experiment, will not tell us anything useful about the initial rate of the reaction but will only tell us the average rate over the time that the reaction is allowed to run for. Since this time will be different for each pH, the results will be meaningless. **(4)**

2 In experiment 1, as the protein is broken down the cloudiness reduces and the absorbance reading falls. In the second case, the initially clear solution becomes cloudy as starch, which is insoluble in water, is formed. **(3)**

17. Investigating enzyme activity

1 (a) absorbance reading / substrate concentration **(1)**

(b) Variables that might change the initial rate that are not temperature. The important one here would be pH, which would be achieved using a buffer solution. Other conditions which would be relevant are enzyme and substrate concentration, which are already maintained in the protocol. **(3)**

(c) A graph is plotted of time on the x-axis and the absorbance reading on the y-axis. The gradient of the straight line portion of the graph is then calculated. This will be the initial rate in absorbance units per second. **(3)**

2 In a thermostatically controlled water bath which is regularly monitored by taking temperature readings with a thermometer **(2)**

18. Factors affecting enzymes

1 Graph B **(1)**

2 The solid line shows us that both acid and alkali pHs affect the enzyme to make it less efficient. But it is noticed that the activity is the same at pH 7 for both methods.
This suggests that the acid conditions do affect the enzyme but that after 5 minutes the enzyme is back to normal, and that alkaline conditions affect it irreversibly. **(3)**

3 As substrate concentration [S] increases the initial rate also increases. This is because there are far more active sites available than are needed to deal with low concentrations of substrate. Enzyme molecules are effectively idle at low concentrations. So when the concentration is increased, there are plenty of active sites to deal with the extra substrate. When [S] reaches a certain value, however, every enzyme molecule is working as fast as it can so adding more [S] has no effect on initial rate. The enzyme is said to be saturated with substrate. **(4)**

19. Enzyme inhibition

1 (a) competitive inhibitor
The malonic acid blocks the active site of the enzyme because it is similar in shape to succinic acid. This slows the enzyme down so the reaction takes longer to reach the plateau with malonic acid than without. However, it does not stop the reaction so it can still reach the plateau. **(3)**

(b)

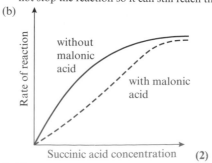

2 **B (1)**

20. Water and ions

1 *Credit is obtained for deploying your knowledge and understanding of the topic. The bulleted points below are called indicative content and as such you do not have to make all the points listed but you can make points which are not listed.
   - clear statement of dependent variable, i.e. exactly what is to be measured stated: mass of plant tissue, mass of fruit, length of shoot, number / colour of leaves
   - clear description of method of measuring change in dependent variable
   - clear statement of independent variable = concentration of calcium
   - range of suitable concentrations suggested (at least five)
   - some clear consideration of time period over which the growth will be measured
   - identification of other variables that could affect growth
   - description of how identified variables can be controlled
   - idea of need for replica at each concentration
   - control of source of plant, e.g. use of same species / variety / source of seeds
   - use of graph to identify other values of concentration to test to identify optimum concentration

   For example,

   I will be using tomato plants.

   In this investigation, the dependent variable is the number of leaves produced over a known time period. The number of leaves will be counted every week for a period of 10 weeks. The tomato plants will be grown in pots containing soil-less compost to which is added $100\,cm^3$ of a 1% calcium carbonate solution once a week. Distilled water is used to water the plants as and when it is needed. The plants will be grown in a controlled environment chamber in which the temperature is kept at 20 °C and light is provided by Gro-light tubes which are on for 12 hours per day. The tomato plants will be of the same variety and all grown from the same packet of seeds.

   The experiment will be repeated at concentrations of 2%, 3%, 4% and 5% calcium carbonate solution.

   The results will be plotted on a graph: x-axis calcium carbonate concentration, y-axis number of leaves. If the graph does not level off, higher concentrations of calcium carbonate will be tested until the point at which the growth levels off is found. This point shows the optimum concentration of calcium to be used. Beyond this concentration it would be uneconomic to add more calcium and achieve no better growth. **(6)**

2 **(5)**

Feature	Example of importance to living things
high specific heat capacity	keeps the temperature in water bodies fairly constant from season to season
polar solvent	allows the transport of nearly all biologically important substances
surface tension	there is a skin over water which helps organisms move on its surface and in plant transport
incompressibility	used in hydraulic systems in living things like hydro skeletons in starfish
maximum density at 4 °C	allows aquatic organisms to carry on with their life even when temperatures are below freezing

21. Exam skills

1 (a) The enzyme increases the rate of reaction, which is higher than the rate if no enzyme is present at all values of substrate concentration [S]. The rate of reaction with the enzyme present is non-linearly related to [S] whereas without enzyme the trend is linear. The increase in initial rate of reaction is the same with or without enzyme present above [S] of $11\,mg\,cm^{-3}$. **(3)**

   (b) (i) ester **(1)**
      (ii) glycerol and fatty acids **(2)**
      (iii) The pH would be reduced due to the production of fatty acids which would ionise to give rise to protons. **(3)**

22. Exam skills

1 (a) A is adenine; C is cytosine; G is guanine; T is thymine **(1)**
   (b) Triplet means the bases on three nucleotides code for each amino acid. In this case, 12 nucleotides code for four amino acids. For example, AAT codes for leucine. Non-overlapping means that each triplet is discrete. So in this case, AAT, AAC, CAG and TTT give four separate codes.
   Degenerate means that more than one code can be used for a particular amino acid. In this case, AAT and AAC both code for leucine. **(3)**
   (c) **B (1)**
   (d) A strand of mRNA with sequence UUA UUG GUC AAA would be formed. This would bind to a ribosome that is involved in protein synthesis. tRNA molecules attached to one specific amino acid would then bind to the codon on the mRNA which is complementary to their anticodon. For example, in this case, a tRNA carrying the amino acid valine would have anticodon GUC and would pair with the CAG region of the mRNA. The two are held together by hydrogen bonds. This would also happen on the next codon with another tRNA / amino acid pairing. Finally, peptide bonds are formed between the two amino acids on the tRNAs by enzymes associated with the ribosome to make part of a polypeptide chain. **(5)**

23. The cell theory

1 (a) **D (1)**
   (b) E, D, F, C, digestive **(2)**
   (c) **C (1)**
2 New cells are formed from other existing cells. **(1)**

24. Prokaryotes

1 (a)

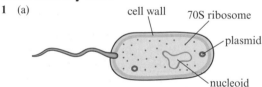

   two marks for ribosome, one for 70S and one for ribosome **(5)**
   (b) yes, because the wall has only one layer, which means the cell is of a Gram positive kind **(2)**
2 Gram negative, B; Gram positive, A; both types, C **(2)**
3 They contain genes that aid the bacterium's survival, such as antibiotic resistance or toxin-producing genes. **(3)**

25. Eukaryotes

1 **(8)**

Description	Name	Function
a large organelle with an envelope with pores through it	nucleus	stores DNA
a branching series of channels studded with small, roughly spherical, structures	rough endoplasmic reticulum	synthesis and transport of protein
quite large oval organelles with folded membranes inside	mitochondria	the site of respiration
a pair of cylindrical structures at right angles to each other	centriole	make the spindle fibre in cell division

2 (5)

Animal cell only	Plant cell only	Animal and plant	Bacteria only	All three cell types
centrioles		mitochondria smooth endoplasmic reticulum (SER) DNA in a nucleus		ribosomes cell surface membrane

3 Plant cells contain some organelles not found in animals. These include chloroplasts where photosynthesis occurs, a vacuole where water and minerals are stored. Outside the plant cell surface membrane is a cell wall. **(4)**

26. Microscopy

1 Electron – Advantage: resolution is very high;
Disadvantage: living material cannot be viewed
Light – Advantage: live specimens can be viewed;
Disadvantage: resolution is very limited **(4)**
2 Bacteria are too small to be seen and the staining might not show them up anyway. **(2)**
3 Generic will be:
   • correct measurement of cell length on the picture
   • converts mm to µm by multiplying by 1000
   • shows division by magnification
   • rounds answer to one decimal place **(4)**

27. Practical microscopy

1 (a) Lignin is only seen in xylem cells at the top part of the plant but in the middle there is more lignin and it is starting to be found in the fibres, although not as much as in the xylem. At the base, lignin is found in both xylem and fibre cells in equal amounts. **(3)**
   (b) diameter of top part is 42 mm on this page magnification is ×52 so actual diameter is 42/52 mm = 0.81 mm
   diameter of base is 45 mm on this page so actual diameter = 0.87 mm
   so increase in diameter is 0.06 mm **(3)**

28. Use of the light microscope

1 (a) A is stage micrometer; B is eyepiece graticule **(2)**
   (b) nucleus is three small units and cell is 15 small units, so cell is $15 \div 3 = 5$ times longer then nucleus **(2)**
   (c) 21 eyepiece graticule divisions correspond to 0.1 mm on the stage micrometer
   One eyepiece graticule division must correspond to $0.1/21 = 0.0048$ mm (2 s.f.)
   one cheek cell is 15 units long on the eyepiece micrometer
   one eyepiece micrometer division is 0.0048 mm long so the length of this cell must be $15 \times 0.0048 = 0.072$ mm (2 s.f.) long **(2)**

2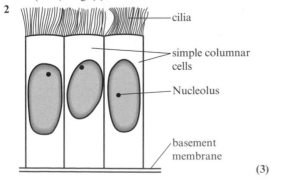

cilia

simple columnar cells

Nucleolus

basement membrane

(3)

29. Viruses: classification

1 Because there are two types of RNA viruses – the RNA viruses proper and the retroviruses. RNA viruses have RNA as their genetic material and make new RNA from it. The retroviruses also have RNA as their genetic material but make DNA from it using reverse transcriptase. Ebola is an RNA virus, HIV is a retrovirus. DNA viruses have DNA as their genetic material, an example is lambda phage. **(4)**
2 DNA virus is C; RNA virus is A; retrovirus is B **(2)**
3 C **(1)**
4 Reproduction, but this is carried out by the host so not strictly speaking a characteristic of the virus itself. **(2)**

30. Viruses: replication

1 New RNA virus will be made by the host cell and then enzymes will make viral proteins using host cell amino acids. Virus particles are assembled inside cells.
There is a delay of 24 hours but in that time the virus is replicating very fast and invading more host cells. **(5)**
2 target the receptors by which viruses recognise their host cells, A; target integrase, D; target protease enzymes, C; target reverse transcriptase, B **(2)**

31. Ebola

1 B **(1)**
2 *Since the disease has an incubation period of 2–21 days any people who are in contact with victims need to be isolated for a minimum of three weeks to ensure people who are incubating the disease with no symptoms do not spread the disease. Due to the droplet infection risk, every effort must be taken to prevent people coming into contact with contaminated body fluids. Victims and contacts must be cared for by people wearing masks and ideally in facilities that prevent air escaping – maybe in a sealed room – which has its air treated and recycled. Hygiene is crucial due to the diarrhoea and bleeding. This will involve good sanitation and attention to hand washing. Preparation of bodies for burial must be carried out in a way that minimises the chances of contact. This may involve the wearing of gloves and a facemask. **(6)**

32. Ebola drug development

1 Double blind trials are used so as not to introduce bias. The sample size is large so that statistical tests can be done to show the significance of any effects seen. More often than not, control groups will be included to allow comparisons to be made. **(3)**
2 Tested on animals to check for safety and toxicity. Then tested on humans to look for side effects and efficacy. Tested on both so metabolism of the drug in whole organisms as compared to cell or tissue culture can be studied. **(2)**
3 (a) A vaccine prevents disease; a treatment cures a disease once someone has it. **(2)**
   (b) It would be good to have both but in the long term it is much better to have a vaccine so that the whole population can be protected from the effects of a disease. Prevention is always to be preferred over cure. If a population is vaccinated against a disease, no-one will get that disease. If a population is not vaccinated against a disease and there is an outbreak of infection, many people may be affected. Treatment may cure them all but it can only be given once someone has the symptoms of the disease. **(3)**

33. The cell cycle

1 The chromosomes condense and therefore become visible as two chromatids. The nuclear membrane breaks down. Centrioles move to opposite end of the cell and form spindles out of microtubules between the centrioles. The chromosomes line up on the equator attached to the centrioles by spindle fibres from their centromeres. The centromeres now split giving rise to two separate chromatids which are pulled apart by contraction of the spindle fibres. When they reach the poles of the cell, anaphase is over. **(5)**
2 (a) A is metaphase; B is anaphase **(2)**
   (b) This cell is in interphase and it is possible to tell this because the chromosomes are not visible. **(2)**

34. Roles of mitosis

1 (a)

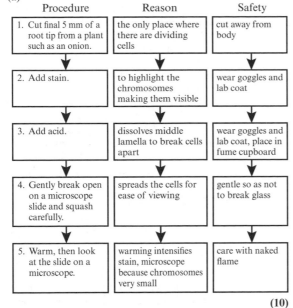

Procedure	Reason	Safety
1. Cut final 5 mm of a root tip from a plant such as an onion.	the only place where there are dividing cells	cut away from body
2. Add stain.	to highlight the chromosomes making them visible	wear goggles and lab coat
3. Add acid.	dissolves middle lamella to break cells apart	wear goggles and lab coat, place in fume cupboard
4. Gently break open on a microscope slide and squash carefully.	spreads the cells for ease of viewing	gentle so as not to break glass
5. Warm, then look at the slide on a microscope.	warming intensifies stain, microscope because chromosomes very small	care with naked flame

(10)

(b) ethanoic (acetic) orcein (Other stains are acceptable, such as toluidine blue or Feulgen stain.) **(1)**

2 (a) 16 **(1)**

(b) mitosis **(1)**

3 Greenfly reproduce asexually in the summer to reproduce faster so they can take advantage of the food and their numbers can increase very rapidly. However, the populations still need to create variation to cope with any potential changes in the environment so they reproduce sexually in the autumn. **(2)**

35. Meiosis

1 A **(1)**

2 **(4)**

Statement	Meiosis	Mitosis
produces variation	✓	
involves two divisions	✓	
used in growth of tissues		✓
used in reproduction	✓	✓

3

The sister chromatids have been pulled apart at the centromeres, which have now split. Two chromosomes are now moving to each pole of each cell. **(4)**

36. Chromosome mutations

1 C **(1)**

2 (a) During anaphase I of meiosis, one pair of homologous chromosomes may fail to separate (non-disjunction) resulting in both chromosomes ending up at the same pole. The gametes formed will have an extra chromosome. At fertilisation, with a gamete having a single set of chromosomes, the zygote formed will have one pair of every chromosome except the one that underwent non-disjunction, of which there will be three chromosomes. **(5)**

(b) Turner's syndrome is an example of monosomy. The difference is that, in this case, the gamete will lack one of the chromosomes; in Turner's this is the X chromosome. When the zygote is formed, the resulting genotype is XO. **(3)**

3 (a) In the first diagram, the chromosomes have swapped parts with each other. This is a balanced mutation. In the second diagram, one chromosome has lost some genes and the other has gained them; this unbalanced. **(4)**

(b) translocation **(1)**

37. Gametogenesis

1 a site containing an enzyme which digests the zona pellucida, C; a site with a haploid number of chromosomes, B; a site containing mitochondria, G; the zona pellucida, F **(4)**

2 **(7)**

Feature	Egg only	Sperm only	Sperm and egg	Neither sperm nor egg
mitochondria			✓	
DNA			✓	
cortical granules	✓			
membrane			✓	
cell wall				✓
diploid nucleus				✓
mid-piece		✓		

3 **(4)**

spermatogenesis	oogenesis
one primary spermatocyte	one primary oocyte
two secondary spermatocytes	one secondary oocyte and one polar body
four spermatids	one ootid and three polar bodies
four sperm	one ovum

38. Plant sexual reproduction

1 pollen mother cell, 20; an endosperm cell, 30; a pollen tube nucleus, 10; a generative nucleus, 10; a polar nucleus, 10 **(5)**

2 C **(1)**

3 (a) The mean pollen tube length increases with time but the mean pollen tube length is always greater with the additive present. This means that the pollen tube growth rate is greater in the presence of the additive than without it. **(3)**

(b) If the pollen tube grows faster it will reach the ovule quicker and then fertilisation will occur quicker. **(2)**

39. Effect of sucrose on pollen tube growth

1 The pollen grains will need to be subject to a range of sucrose concentrations, which is the independent variable. A suitable range would be between 0 M and 2.0 M at 0.4 M intervals. Other factors which might affect germination, such as pH, should be kept at a constant value pre-determined by preliminary work. Similarly, the concentration of mineral ions should be kept the same in each molarity. The total number of grains in a suitable field of view of the microscope would be counted after 60 minutes. The number of these which had germinated should also be counted. This is the dependent variable. **(5)**

2 Pour 250 ml of the 2 M sucrose into a 1 litre volumetric flask. Fill the flask to the 1 litre mark with distilled water. **(2)**

40. Exam skills

1 (a) (i) Enzymes break down the tissue of the style. This will involve the hydrolysis of protein and the middle lamellae between the cells. The breakdown products will supply nutrients and a source of ATP through cellular respiration for pollen tube growth. **(3)**

(ii) egg cell and both polar nuclei **(2)**

(b)

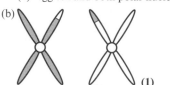

(1)

41. The classification of living things

1 **B (1)**

2 The organism belongs to the Bacteria because it has a peptidoglycan cell wall and is sensitive to antibiotics – neither of these features are shared by Archaea or Eukaryota. **(2)**

3 *The organisms need to be placed in the optimum environment for them to be able to reproduce. There will have to be a mix of mature males and females of both types of organisms in an appropriate number according to their observed mating behaviour. They will need to produce offspring which survive and in due time mature and themselves reproduce to produce viable offspring. **(6)**

42. Molecular phylogeny

1 (a) **B (1)**

(b) percentage change = (22 ÷ 687) × 100;
= 3.2%;
so time taken = 3.2 ÷ 0.2 = 1.6 million years **(3)**

(c) the total number, position and size of the bands could be compared **(3)**

43. The validation of scientific ideas

1 The research is checked for its:
Validity: are its conclusions based on good methods and are the data reliable?
Significance: does it make a useful addition to the existing body of scientific knowledge?
Originality: has someone else already done the same work? This is done by someone respected in the field. **(3)**

2 He could have presented his findings at a scientific conference as a read paper or as a poster display. **(3)**

3 The DNA molecule is the same in all organisms, supporting Darwin's idea of descent from a common ancestor. Estimates of the speed of mutation in DNA have shown that species have evolved over vast periods of time, just as Darwin thought. **(2)**

44. Evolution by natural selection

1 **(2)**

Adaptation	Behavioural	Physiological	Anatomical
production of formic acid as an alarm signal	✓	✓	
acting in a group to carry heavy prey to the nest	✓		

2 Because it is likely that the 1% which are not killed by the product are in some way resistant to it. This means that they will have access to any resources (food etc.) unhindered by their competitors which have been killed by the product. They will quickly reproduce (being bacteria they may double in number as rapidly as once every 30 minutes or so) and pose a bigger threat than before application of the product. **(3)**

3 The GP might say to the patient that prescribing antibiotics is not always the best idea since the probability of bacteria becoming resistant to a particular antibiotic increases if a person does not finish the course because too many then survive and the remaining bacteria may mutate as they reproduce and develop resistance. **(3)**

45. Speciation and the evolutionary race

1 **B (1)**

2 Sympatric speciation is when the populations are isolated in a way other than by geographical separation and this leads to speciation. In this case, reproductive isolation has occurred due to the two distinctly different life-cycle timings. Consequently, the two populations of flies in the wild do not have mature reproductive adults existing at the same time. **(5)**

3 A gene that was involved in the control of lip shape mutated. Also, one coding for height mutated as well. The change in lip shape resulted in better adaptation for feeding. Also, the greater height of white rhinoceros protected it in the open. The alleles for height and lip shape were passed on to offspring. This led to a change in allele frequency. The different food sources resulted in selection pressures. The rhinos would have become reproductively isolated. **(5)**

46. Biodiversity

1 The sharing strategy leads to more species of birds, i.e. more diversity. Therefore, it is better to light log all the land rather than to intensive log some of the land leaving some land unlogged as 'set-a-side'. **(3)**

2 Two plots should be marked out in the same general area. Both should be fenced from sheep to stop sheep from entering or leaving. Sheep are placed in one and the other is mown. After a suitable period of time (say one year), sampling of the plants should be done. Quadrats of suitable size (pre-determined) should be placed randomly in both plots. Plant count should be carried out to give data about both the number of different species and the number of individuals of each species. The results can then be compared using an index of diversity calculation based on the formula:

$$D = \frac{N(N - 1)}{\Sigma n(n - 1)}$$

The figures obtained from this will allow comparison of the species diversity of the two sites. The higher the D, the more diverse the habitat. **(5)**

47. Ex situ conservation: zoos and seed banks

1 Some people think that animals behave unnaturally in zoos. Also, a high proportion of animals kept in zoos are not endangered. Some people feel that animals are kept in poor conditions in at least some zoos. **(3)**

2 *Seed banks are advantageous because the seeds are stored in cool and dry conditions, which means they can be stored for a long time. This is less costly than conserving living plants so large numbers of plants can be stored. Their viability can be tested at regular intervals. The species is less likely to be damaged by natural disasters and disease. **(6)**

3 90% genetic diversity keeps many alleles in the population, thus providing a wide variety of phenotypes. The population needs to show a lot of phenotypic variation as the tiger lives in a wide range of habitats in the wild. Also, if there were a change in the environment, the tigers would be unlikely to be able to adapt if genetic variety was low. **(4)**

48. In situ conservation: habitat protection

1 (a) The reserves and parks will give legal protection to the habitats they contain. The designation will bring international recognition, leading to money for research. In addition, the parks will encourage ecotourism, which will bring in money which can be used for conservation purposes. **(3)**

(b) The buffer zone stretches across the whole of the southern edge of the reserve and therefore separates the centre from the areas outside the reserve. This limits the amount of human activity and conserves the core, protecting it from illegal human activity such as poaching. **(3)**

(c) The government is not well developed and the average income per head shows that Guatemala is a developing country.

It is likely that the government will have other priorities for spending limited money.

The people are poor and likely to take risks, and even commit offences against the laws attempting to protect the reserve and its habitats.

However, world biosphere reserves are internationally recognised and attract funding more easily. **(4)**

49. Exam skills

1 (a) one elephant per 2 km² would be best; that is, a population of 33 000 elephants **(2)**

(b) *In 1980, the elephant population was greater than the 0.5 km⁻² required to achieve maximum biodiversity, so it would be below maximum. From 1980 until 1985, the population declined – as the elephant population fell, so biodiversity would have risen. Just before 1985, however, the population fell below the optimum, leading to a fall in biodiversity again. From the mid-1980s until the early 1990s, biodiversity would remain below maximum, although only slightly. By 1995, probably due to the new law protecting elephants, the population had exceeded the optimum level for biodiversity so it would have started to decline again. It would be likely to reach a minimum for the two decades in 2000 when the elephant population was almost three times higher than the level giving maximum biodiversity. **(6)**

(c) Data on the number of different species would be needed. Various sampling methods, including quadrats, traps, netting and various other techniques, could achieve this. Also, an estimate of the numbers of individuals within each species would be needed. For this, quantitative methods would be needed, such as counts in known areas (e.g. quadrats) or mark/recapture studies. The data could then be substituted into the formula:

$$D = \frac{N(N-1)}{\Sigma n(n-1)}$$

Where D = the index, N = total number of organisms of all species, n = number of organisms of each particular species. **(4)**

50. The cell surface membrane

1 The fatty acids 'tails' of the phospholipids are hydrophobic. This means they orientate themselves away from water, which is found inside and outside cells. The phosphate heads are hydrophilic and can interact with water. So the molecules form bilayers with the hydrophobic tails on the inside and the hydrophilic heads pointing outwards to interact with cytoplasm or fluid bathing cells (both water based). **(4)**

2 size of image (that is distance from A to B) = (approx.) 6 mm
so real size of object = 6 ÷ 1 000 000 mm = 0.000 006 mm; that is 6 × 10⁻⁹ m or 6 nm **(3)**

3 A, fatty acid tail; B, hydrophilic head; C, channel protein; D, glycoprotein **(4)**

51. Membrane permeability

1 (a) blue **(1)**

(b) biotic: surface area / volume (of beetroot); part of beetroot used; age of beetroot used; variety of beetroot used
abiotic: storage conditions; temperature; wavelength / filter **(4)**

(c) The cell membranes would have been damaged by cutting up of pieces. As a result pigment could leak out of the vacuoles of the cells. **(2)**

(d) Increased ethanol concentrations increase intensity of the pigment. This is because of disruption of the membrane because ethanol is a non–polar solvent which will dissolve lipids. The increase in ethanol concentration causes the solution to be less polar and the orientation of phospholipids depends on water around them. **(4)**

52. Passive movement across membranes

1 There is a clear positive relationship between solubility in oil and ability to cross the membrane. This supports the part of the fluid mosaic model that says that the membrane is made of a phospholipid bilayer – the centre of which is hydrophobic. There is no support from these data for the aspect of the model which suggests that the bilayer is studded with proteins, some of which form channels. There is no obvious correlation between size and permeability. **(4)**

2 **(4)**

Process	Requires energy from respiration (ATP)	Requires a concentration gradient
passive diffusion	incorrect	correct
facilitated diffusion	incorrect	correct
osmosis	incorrect	correct
active transport	correct	incorrect

53. Osmosis and water potential

1 (a) A: The vacuole has the red pigment; there is none in the rest of the cell. This tells us that the solution the tissue has been bathed in is more concentrated than the solution in the cell – we say it is hypertonic to the cell solution. Water moves from the cell to the outside solution by osmosis whilst the red pigment cannot move. This can occur because the cell membrane is selectively permeable.
B: Same but much less red pigment in this cell originally in the vacuole.
C: Same but no red pigment in this cell. **(4)**

(b) (i) zero **(1)**

(ii) Turgor pressure is the force applied by the stretched cell wall on the cell contents as they push on it, due to being full of water. As this cell has lost so much water, the contents are not pushing on the wall so the wall is not pushing back. **(3)**

(iii) plasmolysed **(1)**

54. Active transport

1 Both processes involve the use of vesicles to move contents in bulk transport. Both processes also require energy in the form of ATP. Endocytosis moves substances into the cell but exocytosis transports substances out of the cell. **(3)**

2 no specific channel, Graph A; via a channel protein with no competitor for the movement, Graph B; via a channel protein with a competitor, Graph C
The rate in A increases linearly with the increase in concentration since this is increasing the concentration gradient.
In B with a carrier, the rate levels off as the carrier molecules get saturated with transported molecules. At the plateau, every carrier is in use all the time so adding more transported substance will not increase the rate.
In C, the competitor will sometimes be transported so the rate levels off at a higher concentration of the transported substance, which saturates the carrier molecules. **(5)**

55. Surface area to volume ratio

1 (a) **(5)**

Length of side / cm	Area of surface of cube / cm²	Volume of cube / cm³	$\frac{SA}{V}$ ratio	Time for whole block to become coloured / seconds
13	1014	2197	0.46	380
10	600	1000	0.60	300
7	294	343	0.86	100
5	150	125	1.20	53
3	54	27	2.00	20

(b)

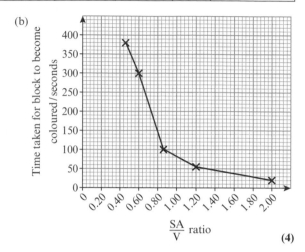

$\frac{SA}{V}$ ratio **(4)**

(c) The graph shows that as the surface area to volume ratio increases, the time decreases for the block to become coloured throughout. Small organisms have a large surface area to volume ratio and the graph shows that such a situation leads to substances reaching the centre of these small objects, blocks or organisms, much more quickly than the larger ones. So, smaller organisms do not need specially adapted gas exchange surfaces such as gills or lungs. **(4)**

56. Gas exchange in mammals

1 The main problem is the reduced diffusion of oxygen and carbon dioxide and therefore less gas exchange. The diagrams show a reduced number and, therefore, surface area of alveoli. There are also fewer capillaries – again leading to reduced surface area and also less blood flow. The latter will reduce the concentration gradient of oxygen from the alveoli to the blood and therefore less oxygen carriage by the blood. **(5)**

2 The number of alveoli in the lung is given by:

$$\frac{\text{volume of lungs in mm}^3}{\text{volume of one alveolus in mm}^3}$$

volume of an alveolus = $\frac{4}{3} \times \pi \times 0.125^3 = 0.008\,\text{mm}^3$
therefore, number of alveoli in two lungs = $6\,000\,000 \div 0.008 = 750\,000\,000$
area of the two lungs is $2 \times 4 \times \pi \times 44.5^2 = 199\,076\,\text{mm}^2$
area of one alveolus = $4 \times \pi \times 0.125^2 = 0.196\,\text{mm}^2$
so, the total area of the alveoli which would fit in two lungs is $750\,000\,000 \times 0.196 = 147\,000\,000\,\text{mm}^3$
so, number of times surface area of lungs is increased by having alveoli is given by the surface area (SA) of all alveoli that would fit in lungs = $147\,000\,000\,\text{mm}^3 \div$ SA of lungs = $199\,076\,\text{mm}^3$
= 738.4 times **(4)**

57. Gas exchange in insects

1 C **(1)**
2 **(5)**

Feature	Advantage
air sacs	act as air reservoirs, increasing the volume of air moved through the respiratory system
chitinous hoops in tracheae	support the tracheae, keeping them open during movement
closable spiracles in body segments	allow the insect to seal its respiratory system off from the air to conserve water
abdominal pumping movements	pump air into the system increasing oxygen supply to active tissues.
non-chitinous tracheoles	allow gas exchange as chitin is impermeable to gases

3 (a) A, spiracle; B, trachea; C, tracheole **(2)**
  (b) Oxygen from the air diffuses into the insect's body cavity through the spiracles, and into long thin tubes called tracheae. The tracheae branch into smaller tubes called tracheoles that have an open ending inside the insect cell, filled with tracheal fluid. Oxygen diffuses into this fluid, and into the insect's cells. **(3)**

58. Gas exchange in fish

1 In the countercurrent exchange system, the blood flowing into the lamella is meeting water that has not yet lost any oxygen. Oxygen will be extracted from the water until the blood is as saturated as the incoming water. In the parallel exchange system, diffusion occurs from the water into the blood until equilibrium is achieved. This equilibrium is at an oxygen concentration halfway between that of the incoming water and the incoming blood. Counterflow is a much more efficient system for the oxygenation of the blood. **(3)**
2 Out of water there is nothing to support the filaments or lamellae, the gills collapse and this leads to a huge reduction in surface area for gas exchange. **(2)**

59. Gas exchange in plants

1 *There is a positive correlation between both K^+ and sucrose with stomatal opening. Both rise as stomata open. However, K^+ rises first followed by sucrose. This suggests that K^+ is transported into the guard cells at dawn. This causes a decrease in water potential (WP) inside the guard cells and water enters down a water potential gradient by osmosis. Guard cells become more turgid and change from a straight (sausage) to a bent cylindrical (banana) shape, opening the pore. Once the pore is open carbon dioxide can enter. This fuels photosynthesis and the manufacture of glucose and then sucrose, which now begins to increase in concentration. K^+ is now moved back out of the guard cells, low WP being maintained by the build-up of sucrose. **(6)**
2 Both allow movement of gases to and from living cells in plants by providing a permeable route through an otherwise impermeable layer. Whereas stomatal permeability can be varied by shape changes in the guard cells opening and closing them as demand changes, lenticels are always open. Guard cells are present in leaves and green stems, lenticels in woody stems. **(3)**

60. The heart and blood vessels

1 *The heart has four chambers: the atria above (in humans) and the ventricles below.
The left and right sides are separated by a septum.
The walls of all chambers are composed of cardiac muscle.
The walls of the ventricles are thicker than those of the atria; the left ventricle being thicker than the right.
The atrioventricular (AV) valves separate the atria from the ventricles and semilunar valves are at the entrance to the arteries leaving the heart.

Tendons are found between the walls of the ventricles and the flaps of the AV valves. The aorta leaves from the left ventricle and the pulmonary artery from the right ventricle. The vena cava and pulmonary vein enter the atria. The heart is supplied with coronary arteries. The pacemaker is found in the right atrium and the atrioventricular (AV) node between the atria and ventricles. Purkyne fibres radiate out from the apex of the heart. **(6)**

2 (a) to prevent blood flowing back again **(1)**

(b) Valves are found in veins, in the pulmonary artery, the aorta and between the atria and ventricles.

They are present in veins because there is not a continuous high pressure pushing blood forward; blood is squeezed forward by body muscles and would fall back when these relaxed again.

They are found in the pulmonary artery and the aorta to stop blood flowing back into the ventricles when they relax after pushing the blood out of the heart.

They are found between atria and ventricles because the blood would move back up into the atria when the powerful ventricle muscles contracted when pushing blood out of the heart. **(4)**

61. Single and double circulation

1 Mammals need more oxygen than fish because they have to move around without the support of water and maintain a constant body temperature.

This takes a lot of energy, so their cells need plenty of oxygen and glucose and produce a lot of waste. Mammals have a double circulation. One part carries oxygenated blood from the heart to the body. The other part carries deoxygenated blood from the heart to the lungs to be oxygenated and carries the oxygenated blood back to the heart. Blood is delivered to the body at high pressure. The blood going through the tiny blood vessels in the lungs is at relatively low pressure so it does not damage the vessels and allows gas exchange to take place. If this oxygenated blood at low pressure went straight into the big vessels that carry it around the body it would move very slowly.

Because it returns to the heart, the oxygenated blood can be pumped hard and sent around the body at high pressure. **(4)**

2 With four separate chambers in the hearts of birds and mammals there can be different pressures produced. The walls of the top two chambers are thinner than the walls of the bottom two, and the wall of the left bottom chamber is much thicker than the wall of the right bottom chamber. Since these walls are made of muscles they are capable of producing pressure and the top two chambers produce lower pressures than the bottom two, and the left bottom much greater than the right one.

The top chambers supply blood to the lower chambers and need very little pressure; the right bottom chamber supplies the lungs at a lower pressure and the bottom left supplies blood to the body at a much higher pressure. **(4)**

62. The cardiac cycle and the heartbeat

1 The sinoatrial node (SAN) is myogenic. Electrical activity from the SAN causes atria to contract. The intrinsic rate can be modified by nerve impulses from accelerator and decelerator nerves. For example, more impulses from accelerator nerves increases heart rate. **(3)**

2 (a) It takes 0.8 seconds for one cycle, so in 60 seconds there will be $60 \div 0.8$ beats.
That is, 75 bpm. **(2)**

(b) AV valve closes, 0.13 s; aortic valve closes, 0.32 s; AV valve opens, 0.5 s **(3)**

(c) During QRS the ventricles are depolarising; during T they are repolarising. **(2)**

63. Blood and tissue fluid

1 **(4)**

	Blood plasma	Tissue fluid	Lymph
Where found	all blood vessels	outside tissue cells and blood capillaries	in the lymph vessels
Protein content	high	low to none	low to none
Glucose content	stable within narrow limits	variable	low
Cell content	red and white	white	white

2 (a) for A, net pressure = $(30 + 6) - (0 + 28) = 36 - 28 = 8$ mm out
in Pa, $8 \times 133.3 = 1066.4$ Pa
for B, net pressure = $(15 + 6) - (1 + 28) = 21 - 29 = -8$ mm (negative, so this is pressure in)
in Pa, -1066.4 Pa **(3)**

(b)

arterial end of capillary   venous end of capillary

HP (30 mm)   OP (6 mm)   blood flow   OP (6 mm)   HP (15 mm)

HP (0 mm)   OP (28 mm)   cells in tissue   OP (28 mm)   HP (1 mm)

A   B

lymph vessel uptake

lymph flow

Key:
HP = hydrostatic pressure
OP = oncotic pressure
(1)

(c) It will lead to an accumulation of tissue fluid in the tissues spaces, called oedema. **(1)**

64. Blood clotting and atherosclerosis

1 A blood clot may form when a blood vessel wall becomes damaged. Cell fragments called platelets stick to the wall of the damaged blood vessel, forming a plug. A series of chemical changes occur in the blood, resulting in prothrombin being converted into thrombin. Thrombin is an enzyme that catalyses the conversion of fibrinogen into long insoluble strands of fibrin. These strands form a mesh that traps blood cells, particularly red blood cells and platelets, to form the clot. **(5)**

2 Endothelial cells become damaged, maybe due to inflammation. White blood cells accumulate in the damaged area. There is also a build-up of cholesterol and fibrous tissue and a plaque forms. This leads to a loss of elasticity of the artery and a narrowing of the lumen. Increased likelihood of clot forming on roughened surface. **(4)**

3 The atheroma narrows arteries, which means there is reduced blood flow to cardiac muscle through the coronary artery. This means that the heart receives less oxygen and starts to respire anaerobically. A by-product of anaerobic respiration is lactic acid, which builds up and causes pain. **(4)**

65. Haemoglobin and myoglobin

1 (a) at pCO_2 20 mm Hg, 20 Hb is 87% saturated at pO_2 of 40 mm Hg
at pCO_2 80 mm Hg, 80 Hb is 58% saturated at pO_2 of 40 mm Hg
so, Hb carries $87 - 58\%$ less $O_2 = 29\%$ **(3)**

(b) Bohr effect **(1)**

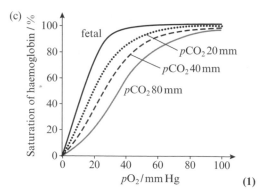

(c) (1)

66. Transport of water in plants

1 (a) (i) A **(1)**

(ii) F **(1)**

(b) E has long, dead, hollow cells with no end walls to allow the movement of water.

F has tubular cells but the end walls are perforated but present. The cells are alive, which gives energy to help the movement of sugars and amino acids in solution. Both are involved in transport of materials around the plant. **(5)**

2 In the centre of the stem, but before the xylem, lies the endodermis. This has a waterproof layer called the Casparian strip. This means that water and anything in solution in it is forced through the living cells and thus the surface membrane. These membranes are not permeable to charged ions such as sodium and chloride. **(3)**

3

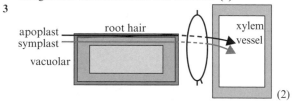

(2)

67. The cohesion-tension model

1 Cavitation is much more common in the day and when wind speed is high. Light causes stomatal opening for gas exchange for photosynthesis and thus increases water loss. This in turn will put more tension on the water columns, pulling more water up from the soil. The water columns in the xylem are more likely to break as a result. Similarly, higher wind speed will lead to greater rate of transpiration as water evaporates from the surface of the leaves. This will increase water movement up the trunk and, in the same way, increase tension and thus cavitation. **(4)**

2 Both refer to interactions between molecules. Cohesion describes the force of attraction between like molecules (for example between water molecules). Adhesion describes the force of attraction between unlike molecules such as water and the walls of xylem vessels or capillary sized pores in mesophyll cell walls. **(4)**

68. The mass-flow hypothesis

1 The concentration of exudate is highest at the top of the tree in summer as this is where the leaves are found and they are actively making sugar at this time of year.
Mass flow needs such a difference and would therefore account for movement down the tree from leaves, source, to roots, sink. In the autumn, there is no such gradient and there is no flow in the phloem either; as there are no longer any leaves, the source and sink have disappeared. **(4)**

2 The mass-flow hypothesis has no role for living cells; the process can happen with no need for metabolism. The statements which argue against this idea are those which talk about living companion cells and the differential rate of movement of different sugars together with the possibility of bi-directional movement. On the other hand, sap exudation from aphid mouthparts and the evidence for phloem loading at leaves supports a mass-flow idea. **(5)**

69. The uptake and loss of water

1 This is because water uptake and water loss are related. Nearly all of the water taken up is lost through the leaves because only a tiny amount is used by the plant in photosynthesis. Some water is produced by the plant respiration, but again the amount is tiny and is used for photosynthesis, further reducing the amount of water taken from the soil that is actually used in metabolism. **(2)**

2 (a) temperature, relative humidity, air movement **(3)**

(b) temperature: if this was not constant, transpiration rates would change, getting faster as it got warmer or getting slower as it became cooler
relative humidity: if this was not constant, transpiration rates would change, getting faster as it became drier or getting slower as the air became more moist
air movement: if this was not constant, transpiration rates would change, getting faster as it became windier or slower as it became more still **(2)**

(c) It is generally observed that in both cases as light intensity increases so does the rate of transpiration. Again in both cases, the rate of transpiration levels off as light intensity reaches about 0.3 arbitrary units. At all light intensities, the rate from the upper surface of the leaf is less than that from the lower which could be due to a combination of a thicker cuticle on the upper surface together with the stomata. **(3)**

70. Exam skills

1 (a) Cohesion is the force that holds molecules of the same kind together. In this instance, they are water molecules and they are held together by hydrogen bonds. **(2)**

(b) (i)

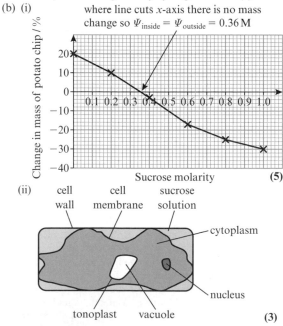

where line cuts x-axis there is no mass change so $\Psi_{inside} = \Psi_{outside} = 0.36\,M$ **(5)**

(ii) (3)

71. Exam skills

1 (a) Both sets of atrioventricular valves are open, showing that the ventricles are not contracting and blood is flowing through them from the atria to the ventricles. The lumina (lumen) of the ventricles look open and distended, showing they are full of blood. This evidence suggests the heart is in diastole – it is relaxed and filling with blood **(3)**

(b) This condition changes the electrical activity of the heart. The peak is reversed and is earlier than expected. There is no change in pressure in pulmonary artery because there is little blood in the ventricles so when the ventricles contract, little blood is forced into the pulmonary arteries. The normal wave after E is missing, leading to a longer gap before next wave of excitation. As a result, after E there is a long gap before the next normal contraction of the heart and so to the next normal increase in pressure in the pulmonary artery. **(3)**

72. Stages of aerobic respiration

1 (a) (7)

	Glycolysis	Link reaction	Krebs cycle	Oxidative phosphorylation
ATP produced	✓		✓	✓
ATP hydrolysed	✓			
CO_2 produced		✓	✓	
NAD reduced	✓	✓	✓	
Reduced FAD is oxidised				✓
Occurs in cytoplasm	✓			
Occurs in mitochondria		✓	✓	✓

(b) 2ATP is broken down into $2ADP + 2P_i$ + energy; two phosphate groups are added to glucose molecule; by enzymes **(3)**

(c) Hydrogen is supplied to the electron transport chain in the form of reduced NAD

- from glycolysis

- the hydrogen ions and electrons entering the ETC allow oxidative phosphorylation to occur **(3)**

73. Glycolysis

1 (a) A, hexose bisphosphate; B, glycerate-3-phosphate (GP) **(2)**

(b) ATP production in glycolysis, substrate level phosphorylation; addition of phosphate, phosphorylation; removal of hydrogen atoms, dehydrogenation; addition of hydrogen atoms to NAD, reduction **(4)**

(c) glucose is a stable molecule; adding phosphate groups increases its reactivity / destabilises it; the phosphate is used to phosphorylate ADP; reforming ATP **(3)**

74. Link reaction and Krebs cycle

1 (a) mitochondrial matrix **(1)**

(b) CO_2; reduced NAD; reduced FAD; ATP **(2)**

(c) Acetyl CoA is a 2-carbon molecule. It binds to the 4-carbon molecule C to form molecule A which is a 6-carbon molecule. Decarboxylation (and dehydrogenation) occur to form a 5-carbon molecule B. Decarboxylation (and dehydrogenation) occur to form a 4-carbon molecule C. **(4)**

75. Oxidative phosphorylation

1 (a) oxidised NAD/NAD^+ **(1)**

(b) A **(1)**

(c) Electrons pass from carrier A to B to C in a series of redox reactions. This releases energy which is used to pump the (H^+ / hydrogen ions / protons) into the intermembrane space / space between the inner and outer mitochondrial membrane. The structure X is ATP synthase. H^+ flows through X down a concentration gradient / electrochemical gradient. Sufficient energy is released to phosphorylate ADP to form ATP. **(5)**

76. Anaerobic respiration

1 (a) A **(1)**

(b) Molecule S is pyruvate and it is reduced by the addition of hydrogen atoms to form lactate. These hydrogen atoms come from reduced NAD. Oxidised NAD is produced. Glycolysis can continue, producing some ATP. **(3)**

(c) Three points from:
- Both recycle oxidised NAD to allow more glycolysis to occur.

- In animal cells, pyruvate is reduced whereas in plant cells ethanal is reduced.
- Animal cells produce lactate whereas plant cells produce ethanol/alcohol.
- Carbon dioxide is not produced in animal cells whereas it is in plant cells. **(3)**

77. Using a respirometer

1 (a) potassium hydroxide / soda lime **(1)**

(b) *The apparatus should be placed in a water bath at 25 °C and left to equilibrate; mass / volume of respiring peas; mark the level of the dye at start and end of the experiment; measure the distance moved by the dye; calculate volume of oxygen taken up by peas using $\pi r^2 \times$ distance; replicates; calculation of mean; calculation of rate; use the syringe to reset the dye position between replicates **(6)**

(c) $\pi r^2 \times$ distance
$\pi \times (1.5)^2 \times 52 = 367.5 \, mm^3$
ans $\div 3 = 122.5 \, mm^3$ / min **(2)**

78. Photosynthetic pigments

1 (a) in photosystems in the thylakoid membranes **(1)**

(b) each pigment absorbs light energy of different wavelengths; having more pigments increases the number of wavelengths of light that the plant can absorb; increases plant photosynthesis **(3)**

(c) (i) this is the control tube; to show that only wavelength of light is producing the effect **(2)**

(ii) very low light energy absorbance; little or no photosynthesis occurring; algal cells respiring at a greater rate than photosynthesis; carbon dioxide levels increase causing a decrease in pH **(3)**

79. Investigating photosynthesis

1 (a) Two more possible variables are:
- number of algal balls
- volume of indicator **(1)**

(b) 1.6 **(1)**

(c) Algal balls in the tubes closer to the light source received light at higher intensity. More light energy was absorbed by the photosynthetic pigments in the chloroplasts. Photosynthesis occurred at a faster rate / more photosynthesis occurred. More carbon dioxide was absorbed from the indicator solution. **(2)**

80. Chloroplasts

1 (a) The chloroplast should be between 5 and 10 μm in diameter.
If I gave magnification of 5430 and the image was 3.8 cm wide it would give an actual size of approx 7 μm. **(2)**

(b) (i) C **(1)**

(ii) B **(1)**

(c) (i) starch **(1)**

(ii) Two answers from the following:
- glucose molecules formed in the light-independent stage
- joined together by condensation reactions
- 1,4 **and** 1,6 glycosidic bonds **(2)**

(iii) One more answer from the following:
- The DNA will contain the genetic code for each enzyme.
- Protein synthesis occurs on the ribosomes. **(2)**

81. Photosynthesis: light-dependent stage

1 (a) thylakoid membranes **(1)**

(b) **(7)**

	Cyclic photophosphorylation	Non-cyclic photophosphorylation
Electrons are passed along an electron transport chain	✓	✓
Redox reactions involved	✓	✓
Electrons raised to higher energy level	✓	✓
Electron from a water molecule replaces electron lost from chlorophyll		✓
The same electron replaces electron lost from chlorophyll	✓	
NADP reduced		✓
ATP produced	✓	✓

(c) *Three more points from:
  • less light-independent reactions / Calvin cycle can occur
  • less glucose produced
  • less cellulose / proteins can be produced
  • fewer new cells can be produced by mitosis **(6)**

82. Photosynthesis: light-independent reactions

1 (a) (i) ATP and reduced NADP **(2)**
    (ii) stroma **(1)**
  (b) Carbon fixation occurs when the RuBisCO enzyme catalyses a reaction between carbon dioxide and RuBP. Molecules of GP are formed. Energy from ATP hydrolysis and hydrogen atoms from the oxidation of reduced NADP reduce GP into GALP. **(4)**
  (c) nitrogen; sulfur **(2)**
  (d) - polysaccharide
      - composed of glucose molecules joined together with glycosidic bonds
      - composed of amylose and amylopectin
      - amylose contains 1,4 glycosidic bonds and is not branched
      - amylopectin contains 1,4 & 1,6 glycosidicbonds and is branched **(4)**

83. Limiting factors in photosynthesis

1 *An increase in temperature up to the thermal optimum would increase the rate of the photosynthetic enzymes and therefore increase the rate of photosynthesis.
  Light intensity: an increase in light intensity would mean more light energy could be absorbed by the photosynthetic pigments; more light-dependent and light-independent (LI) reactions could occur.
  Carbon dioxide concentration: an increase in carbon dioxide concentration will increase the amount of carbon that can be fixed in the LI stage and the amount of glucose produced. **(4)**

2 Any five answers from the following:
  • competition between oxygen and carbon dioxide for RuBisCO active site
  • as oxygen (concentration) increases more RuBisCO reacts / binds with oxygen / less RuBisCO reacts / binds with carbon dioxide
  • less RuBP formed / less carbon fixation occurs
  • less GP / GALP produced
  • less glucose / lipids / amino acids produced
  • less plant growth **(5)**

84. Exam skills

1 (a) **(3)**

	Name of reaction pathway	Location in leaf cell	Occurs in photosynthesis	Occurs in respiration
W	glycolysis	cytoplasm		✓
X	Calvin cycle / light independent stage	stroma (of chloroplast)	✓	
Y	Krebs cycle	(mitochondrial) matrix		✓

(b) ATP; water; oxidised NAD/FAD **(2)**
(c) all three are reduced by accepting hydrogen;
  reduced NAD carries hydrogen to the electron transport chain / for oxidative phosphorylation;
  reduced FAD carries hydrogen to the electron transport chain / for oxidative phosphorylation;
  reduced NADP carries hydrogen to the light-independent stage / Calvin cycle **(4)**

85. Culturing microorganisms

1 1 is **B**; 2 is **C**; 3 is **A**. All need to be correct for one mark. **(1)**
2 **D (1)**
3 **C (1)**
4 Three from the following:
  • use aseptic technique to inoculate nutrient agar plates or nutrient broth tubes.
  • controls with no bacteria added
  • incubate at different temperatures ranging from 0 °C to 60 °C
  • measure growth by counting colonies or assessing absorption or transmission using a colorimeter **(3)**

86. Growth of bacteria

1 (a) 40 (No bacteria are on the outer (3 lined) boundaries.) **(1)**
  (b) (i) $40 \div 0.004$
        $= 10\,000$ or 10^4 **(2)**
    (ii) $10^4 \times 10^3 = 10^7$ or 10×10^6 or 1×10^7 **(2)**
    (iii) $10^7 \times 10^6 = 10^{11}$
        $= 10 \times 10^{10}$ or 1×10^{11} **(2)**
  (c) Actual length in mm $\times$ 1000 to give µl then divide by 1000 **(2)**

87. Streak plating and measuring population growth

1 (a) because the range is too large to use a linear scale **(1)**
  (b) any number between 100 and 205 but not 0.5 as you cannot have half a colony **(1)**
  (c) $50 \times 10 = 500$ **(2)**
  (d) 30 minutes **(2)**
  (e) Three from the following:
    • lower numbers / rate of growth
    • at lower temperatures, enzyme rates of reaction are lower
    • less kinetic energy
    • enzyme-catalysed reactions, such as DNA replication / protein synthesis, are slower **(3)**

88. Bacteria may cause diseases

1 C (1)
2 Exotoxins are lipopolysaccharides. (F)
  Exotoxins pass out of bacterial cells. (T)
  Endotoxins are made by Gram negative bacteria. (T)
  Exotoxins can damage red blood cells. (T) (4)
3 (a) Penicillin acts on bacterial cell walls by preventing
    growing Gram positive bacteria from synthesising
    their cell walls. Antibiotics kill the infecting bacteria so
    they cannot produce the exotoxins. Without these, the
    action of the non-specific immune response is stopped
    or reduced greatly and therefore there is no / little
    inflammatory reaction. (3)
    This reduces the risk of organ failure.
  (b) The first symptoms, headache, fever, rash and feeling
    cold, occur with many infectious diseases such as
    influenza / other named infectious diseases. (2)
4 Two from the following:
  • wounds
  • with food into gut
  • droplets breathed into airways
  • in water / faecal-oral route
  • by a vector
  • unsterilised needles (2)

89. Action of antibiotics

1 Two from the following:
  • Human immunodeficiency virus / HI virus, infects, T cells
    / cells of immune system / named cells of immune system,
    e.g. T helper cells.
  • Immune system cannot keep *Mycobacterium tuberculosis*
    bacteria dormant / fewer cells to keep bacteria enclosed
    and inactive.
  • *M. tuberculosis* infects cells in lungs causing damage and
    symptoms. (2)
2 both are made by microorganisms; both reduce competition
  between microorganisms;
  bacteriostatic slow / prevent growth of bacteria whereas
  bactericidal kill bacteria;
  neither are effective against viruses / fungi / eukaryotes (3)
3 (a) • acts on bacterial cell walls (1)
      • acts on Gram positive bacteria that are growing (1)
      • prevents cell wall synthesis so bacteria take up water
        by osmosis, swell and burst/lyse (1)
      • ineffective against Gram negative bacteria as it does
        not inhibit them from making their cell walls (1)
  (b) Two from the following:
      • interferes with protein synthesis / translation
      • binds to 30S subunit of microbial ribosomes
      • bacteriostatic (2)

90. Antibiotic resistance

1 (a) mutation: change to: genetic material / gene / DNA / or
    chromosome;
    hydrolysis: chemical breakdown of large molecule by,
    reaction / action of / addition of, water;
    enzyme: biological catalyst;
    plasmid: small circular piece of DNA in prokaryote /
    bacterial cells (4)
  (b) Each gene has only one allele as bacteria have only
    one single, looped chromosome/do not have paired
    linear chromosomes, (1), therefore any change in a
    gene is expressed / leads to visible characteristics in the
    phenotype. (2)
  (c) shape of its active site is not complementary to the (part
    of the) tetracycline molecule. (1)
  (d) If someone takes antibiotics to kill a pathogen, the
    antibiotics could also kill some of their gut bacteria and
    select resistant bacteria in their gut. These bacteria pass
    out in faeces and when they come into contact with other
    bacteria can pass their resistance genes by horizontal
    gene transfer – they pass copies of their plasmids to other

bacteria of a different species. The gut bacteria have
resistance due to a random mutation and the antibiotic
acts as a selection pressure. Populations of bacteria evolve
antibiotic resistance due to natural selection. (4)

91. Other pathogenic agents

1 C, B, D, F, A, E (All must be in the correct order for two
  marks. One mark can be awarded if three events are placed
  in the correct squence.) (2)
2 Their surface antigens will only fit receptors on certain cells;
  due to their shapes being complementary to the receptors. (2)
3 Two from the following:
  • hand to hand
  • from surfaces someone else with viruses on their hands
    has touched
  • saliva / mucus
  • from pigs / chickens via faeces (2)
4 They are important because they can infect crop plants,
  damaging their cells and tissues and absorbing nutrients from
  the plant. This weakens the plant and reduces its growth,
  therefore reducing its yield for human consumption. This
  leads to loss of profit for growers and less food availability to
  feed the human population. Agrichemicals may have to be
  used to reduce the spread of the virus and this costs money /
  reduces farmer's profits / makes food more expensive. (5)

92. Malaria

1 C (1)
2 (a) Only females suck blood; as they need to ingest extra
    protein, before producing eggs. OR Males do not lay
    eggs; and so do not suck blood. (2)
  (b) People can be without symptoms for 7–10 days after
    being infected as the parasite is inside the liver cells; if
    they stop taking their tablets, when the parasites break
    out of the liver cells and enter the blood they will not be
    killed as there is no prophylactic in the blood. (2)
  (c) Red blood cells are destroyed; therefore loss of
    haemoglobin so less able to carry oxygen / less oxygen
    carried; therefore less respiration and less ATP (for
    muscle contraction/metabolism) (2)

93. Controlling endemic diseases

1 C (1)
2 One of the following sets of points:
  • Spray adult mosquitoes with pesticides to kill female
    mosquitoes; so they cannot be vectors / breed.
  • Release sterile males into the environment so females lay
    infertile eggs; this reduces mosquito population.
  (Answers must be marked in pairs – no mix and match.) (2)
3 *Six from the following:
  • the pesticide acts as a selecting agent / agent of selection
  • some mosquitoes have a random genetic mutation that
    gives them resistance to the insecticide
  • so insecticide / pesticide becomes ineffective
  • need to use a different insecticide
  • insecticides may also kill useful insects
  • removal of those insects (and mosquitoes) reduces food
    sources for animals that eat insects
  • disrupts food chains / webs in ecosystem
  • insecticides can accumulate in food chain and harm top
    predators / humans (6)

94. The role of white blood cells

1 C (1)
2 D (1) (*This is the only correct answer. Although B cells make
  antibodies that are placed in their cell surface membranes and
  act as receptors, they do not shed antibodies into the blood.*)
3 (a) antigen: molecule, usually protein or glycoprotein,
    on cell surface membranes or surface of a virus; that
    stimulates an immune response (2)
  (b) MHC: major histocompatability complex antigen on
    surface of all cells except red blood cells; that tells the

immune system that these cells are 'self' not 'non-self' **(2)**

4 **(6)**

Cells	Stimulate other types of cells involved in specific immune response	Stop the immune response when pathogens are destroyed	Produce antibodies	Contain large amount of rough endoplasmic reticulum	Remain in the body conferring long-term immunity
B cells					
T_h cells	✓				
T_s cells		✓			
T_k cells					
plasma cells			✓	✓	
memory cells					✓

95. The humoral immune response

1 phagocytosis; macrophages; antigen presenting cell; T helper; mitosis; cytokines; mitosis; endocytosis; mitosis; memory; antibodies; antibodies; macrophages; T suppressor; memory; immune **(8)**

96. The cell-mediated immune response

1 T_k cells bind to and destroy pathogens. (F)
Infected host cells present the pathogen's antigens on their cell surface membrane. (T)
The cell-mediated response involves APCs. (T)
T_k cells release chemicals that cause infected cells to lyse. (T)
T_k cells can destroy cancer cells. (T)
T_k cells have receptors that fit the displayed pathogen's antigens and receptors that fit the host cells' MHC antigens. (T)
The cell-mediated response does not involve clonal selection. (F) **(7)**

2 (a) Cell receptors have to have complementary shape; to antigens on the surface of HIV. **(2)**
 (b) Although it has RNA it does not have the enzyme reverse transcriptase. **(1)**
 (c) It was first found in non-human / named animals / fruit bats and then it passed from those animals and infected humans. **(1)**

3 **(2)**

Immune response	B cells involved	T_h cells involved	T_k cells involved	Antigen presentation	MHC involved	Clonal selection	Clonal expansion	Mitosis involved
cell mediated		✓	✓	✓	✓	✓	✓	✓
humoral	✓	✓		✓	✓	✓	✓	✓

97. Types of immunity

1 * **(6)**

	Active immunity	Passive immunity
Natural	person suffers from a disease caused by an infecting agent and mounts an immune response that gives long-term immunity after recovery	baby receives antibodies against antigens of pathogens the mother has been infected with / vaccinated against **(1)**; either before birth across placenta; or after birth in colostrum / first breast milk **(1)**; short lived as child will make antibodies to the received antibodies **(1)**
Artificial	person is vaccinated and their body mounts an immune response **(1)**; antibodies and memory cells made **(1)**; memory cells remain in body for long time **(1)**	antibodies against a specific antigen found on the surface of a particular pathogen are injected into a person to give quick acting but short-lived immunity

2 immunity of most / all people in a population; (so that) pathogen cannot be transmitted **(2)**

3 100% **(1)** (If 100% of the population were to be vaccinated then because the vaccine is only 95% effective, this would give 95% coverage / immunity in the population. In reality, not everyone becomes vaccinated so we rarely achieve herd immunity.)

4 Any three from the following:
 • virus did not mutate
 • so (therefore) its antigens are always the same shape
 • memory cells made after vaccination

- produce antibodies that always fit the smallpox virus antigens **(3)**

98. Exam skills

1 **(3)** (One mark per correct row.)

Microorganism	Not made of cells	Contain nucleic acid	Cells have a nucleus	May produce antibiotics	May be killed or inhibited by antibiotics
viruses	✓	✓			
fungi		✓	✓	✓	
bacteria		✓		✓	✓

2 (a) 60 minutes **(2)**

(b) Any three from:
- solid line shows numbers of bacteria over time
- exponential curve as numbers double every hour
- dotted line shows amount of DNA which doubles as all the DNA in the cells (plasmids and chromosome)
- replicates before each cell divides by binary fission **(3)**

(c) It would decrease / drop / become lower / named number below 6 **(1)**

(d) C **(1)**

99. Using gene sequencing

1 C **(1)** (B refers to gene pool and A is not correct as not all organisms have chromosomes and eukaryotes have DNA in mitochondria and chloroplasts as well as in their chromosomes. Prokaryotes also have a genome so D is incorrect.)

2 **(2)** (One mark per row.)

	Organelle				
Type of cell	Nucleus	Ribosomes	Plasmids	Chloroplasts	Mitochondria
prokaryotic	✗	✗	✓	✗	✗
eukaryotic	✓	✗	✗	✓	✓

*(Note that the question asks **whether** DNA is in these organelles not **whether** the cells have the organelles. However, if the organelles are absent from all prokaryotic or eukaryotic cells then these cells cannot have DNA present in these structures. Ribosomes contain RNA and protein but not DNA. If you change your mind about a response, completely cross the old response out and add a new one. Do not try to make a tick into a cross – it is ambiguous and will be marked as wrong.)*

3 B, D, A, C **(1)**

4 (a) hydrogen (bonds) **(1)** (*This is a synoptic question and tests your knowledge of the structure of DNA. You need to know this really well before studying Topic 7.*)

(b) It is heat stable (Alternative wording: answer does not need to include the words 'heat stable' but can involve a description of what this means - eg. 'not denatured/ changed by heat') as this bacterium survives at very hot temperatures; therefore the temperature does not have to be reduced and then raised, and this speeds up the process. **(2)**

100. Transcription factors

1 C **(1)**

2 (a) *A is double-stranded DNA; B is transcription factor; C is RNA polymerase; D is gene; E is mRNA; F is promoter region **(6)**

(b) B, C (and protein) labelled as protein; the double-stranded DNA, gene, promoter region and E labelled as nucleic acids **(2)**

3 B **(1)**

4 They have a region / DNA-binding domain; that is complementary in shape to the shape of a specific sequence of base pairs so will only bind to that section of DNA. **(2)**

5 1 Development: organisation of tissues and organs in the developing embryo.

2 Hormonal control, e.g. oestrogen passes into a cell through the surface membrane and binds to a receptor in the cytoplasm and this then passes into the nucleus and acts as a transcription factor, activating certain genes.
3 Control of the cell cycle
4 Pathogens change gene expression in host cells so the pathogens can infect the host cell. **(3)**

101. RNA splicing and epigenetics

1 intron is C; exon is A; spliceosome is B; pre-mRNA is D **(4)**

2 C **(1)**

3 Acetylation of histone proteins in chromatin reduces gene expression. If these 'switched off genes' are activated by removing the acetyl groups on the chromatin, then the activation genes can allow transcription of proteins involved in normal brain function and / or memory. The proteins transcribed could be neurotransmitters, receptors, transcription factors for other genes, or enzymes. **(4)**

4 (a) 650 amino acids require 650 × three base triplets = 1950 base pairs
There is also a stop codon so that makes 1953 base pairs for the exons.
introns = (3000 − 1953) bp = 1047 base pairs **(3)**
(*Maximum of two points if the stop codon not taken into consideration.*)

(b) by variations in splicing so that different exons / combinations of exons used **(1)**

102. Stem cells

1 D **(1)**

2 T, F, T, T, F, F, T, F, T, F **(9)**

3 New cells need to be produced often in these tissues; by mitosis / only stem cells can continue to divide (to give rise to new differentiated blood and skin cells). **(2)**

4 Totipotent stem cells are found in very early embryos and can divide and differentiate into any kind of cell (for that organism), including a new embryo; pluripotent stem cells are found in an older embryo / blastocyst and can produce any kind of cell for that organism except an embryo. **(2)**

103. Medical uses of stem cells

1 *iPS cells are made from adult stem cells, such as fibroblasts, that are multipotent and found in skin and other tissues. The fibroblasts are cultured and four genes, called Yamanaka factors, are added to the cells. These genes cause genes that were switched off in the fibroblasts to be activated or switched on again, so that the cells become pluripotent. They have been reprogrammed and can differentiate into almost all types of cell, except an embryonic cell. If the adult stem cells are taken from the person to be treated with the resulting iPS cells, there will be no rejection as the patient's immune system will recognise the iPS cells as self and not foreign. **(6)**

2 (a) **(4)**

	Considerations of medical uses of embryonic stem cells
Benefits	can produce almost all cell types to grow replacement organs and tissues; can be used for testing new drugs (before clinical trials); and this means fewer animals need to be used for drug testing
Disadvantages	ethical considerations; may be rejected by the recipient's immune system, as the organs and tissues have come from another human and will have different antigens on their surface

(b) *Some people have religious views and think that humans have rights from conception, which means embryos have rights; as these embryos being used as a source of stem cells cannot be asked to give consent, this violates their human rights; some people feel that killing an embryo is no different from killing an adult; on the other hand the embryos used for this research / treatments are spares created by IVF and would otherwise be destroyed after a specified period of time; also many people conceive (often without being aware of it) but the embryo fails to embed or develop and is lost; no one can remember being an embryo so the idea embryos having to give consent seems rather bizarre; it also seems rather odd for society to sanction sentient adults with families and responsibilities being able to fight in wars while not allowing this use of spare embryos where the potential benefits are large; ethicists often consider a scenario – if you were in a burning lab would you save the people there or a dish of embryos? **(6)**

104. Recombinant DNA

1 restriction endonucleases, B; reverse transcriptase, D; sticky ends, E; plasmid, A; DNA ligase, C **(5)**

2 (a) liposomes are made of lipid molecules; so even though they are large they can dissolve in and pass through the lipid bilayer of the cell surface / plasma membrane and nuclear envelope **(2)**

   (b) After the first use of the virus as a vector, the recipient mounts an immune response to the virus. When the virus is injected again then memory cells will make antibodies that agglutinate the virus particle together, preventing them from entering the cell. **(3)**

3 (a) sticky ends **(1)**

   (b) gene 1 **(1)**

105. Identifying recombinant cells

1 C **(1)**

2 (a) Press a piece of sterile velvet onto the colonies on the first agar plate; keep the orientation of the velvet the same and then press it onto another sterile agar plate so that some bacteria from each colony on the first plate are placed onto the second plate in the same location (replace lid and incubate plate for 24–48 hours at around 30–35 °C). **(2)**

   (b) A and E
   These colonies can grow in ampicillin agar so they have taken up a plasmid as they have a gene for resistance to ampicillin; but they cannot grow in tetracycline agar as their gene for tetracycline resistance is disrupted; by the human insulin gene that is inserted there. **(3)**

   (c) Each colony originates from one recombinant bacterium; that divides many times (by binary fission) and at each division all the DNA including the plasmid is copied and passed into the daughter cells / all cells in the colony are a clone of the original cell. **(2)**

106. Genetic modification of crops

1 B **(1)**

2 (a) C, E, A, B, D, F, G **(5)**

   (b) Each newly produced genetically modified plant originates from one single cell taken from a gall; that stem cell divides by mitosis and all the DNA is replicated before division so each new cell receives a copy of every gene, including the desired gene. **(2)**

   (c) Selective breeding has caused many genetic changes in crop plants, including changing the numbers of chromosomes. When cross breeding to introduce genes for desired characteristics you cannot be sure which genes are getting into the offspring and sometimes undesirable genes end up in the offspring and some desirable genes may be lost. Too much inbreeding can lead to weaknesses in the plants. Modern methods involve putting only specific genes into crop plants and the results are trialled and tested to see if they adversely affect the environment, whereas selectively bred crop plants were often not tested for their effects on the environment or on human health. **(5)**

107. Exam skills

1 (a) The protein encoded by the *FOXP2* gene binds to special areas on the DNA of other genes; and either activates (upregulates) or represses (downregulates) them; it is a transcription factor that can bind to many different genes and alter their expression so this gene can lead to many different effects in the phenotype of the organism. **(2)**

   (b) Changing the amino acid sequence changes the protein's primary structure, which can change its tertiary structure (as bonds cannot form between specific amino acids in the polypeptide chain); the (DNA-binding domain of the) protein is no longer complementary in shape to the shape of the specific base sequences of the DNA molecule to which it normally binds. **(2)**

   (c) Neanderthals had a version of the *FOXP2* gene that was similar to that of modern humans. In knockout mice with no functional alleles of the *FOXP2* gene, their lungs fail to develop properly; the human *FOXP2* gene is expressed in fetal and adult lung tissue / cells. **(3)**

   (d) autosomal dominant; lack of one functioning allele / presence of one mutant allele leads to DVD **(1)**

108. Sources of genetic variation

1 (a) An explanation that makes reference to any of the following:
   • The genetic code determines the primary structure of a protein.
   • If one base triplet / codon is changed, this may result in a different amino acid being inserted into the polypeptide chain during translation.
   • A change in amino acid sequence means that the polypeptide chain will fold into a different 3D / tertiary structure, changing the role of the protein. **(3)**

   (b) Point mutations may involve insertion, deletion or substitution. The first two result in a frameshift where all the base triplets / codons after the mutation will be altered. This will have a significant effect on the protein, potentially altering many of the amino acids in the primary structure of the protein. Substitutions have less effect as each substitution will only affect one codon. **(3)**

2 An explanation that makes reference to the following:
   For example,
   Down's syndrome;
   non-disjunction of chromosome 21 at meiosis so one gamete has two chromosome 21s;
   resulting zygote has 3 chromosome 21s / trisomy 21 **(3)**
   (*Allow other trisomies such as trisomy 13 or trisomy 18.*)

3 An explanation that makes reference to the following:
   Independent assortment of chromosomes occurs during metaphase I.
   When the homologous pairs of chromosomes line up on the equator of the cell, the orientation of the maternal and paternal member of each pair is random; so that when they are pulled apart into two daughter nuclei during anaphase I, each daughter nucleus contains a mix of (now unpaired) maternal and paternal chromosomes.
   The same process is repeated during meiosis II when the sister chromatids line up on the spindle. The sister chromatids will not be genetically identical as crossing over has occurred during prophase of meiosis I.
   Each chromatid has a different combination of alleles;
   This means the resultant daughter nuclei produced at the end of meiosis II will again be genetically different. **(5)**

109. Understanding genetic terminology

1 genes; genotype / genome; alleles; phenotype; diploid; two; homozygous; heterozygous (NB: last two could be in reverse order, i.e. heterozygous; homozygous) **(8)**

2 cross it with a white flowered individual – if some offspring are white then the purple parent must be heterozygous **(1)**

3 Group A, I^AI^O or I^AI^A; Group AB, I^AI^B; Group O, I^OI^O; Group B I^BI^B or I^BI^O **(4)**

4 fur changes colour in response to exposure to colder temperatures at the extremities; there is a temperature-sensitive enzyme which only functions at cooler temperatures, to produce the pigment melanin **(2)**

110. Genetic crosses and pedigrees

1 offspring genotypes and phenotypes

	I^A	I^O
I^B	I^AI^B Group AB	I^OI^B Group B
I^O	I^AI^O Group A	I^OI^O Group O

1 out of 4 children will have blood group O, so the probability is 1 in 4 or 25% **(4)**

2 because the allele is dominant / expressed in the phenotype even if only one is present, the child only has to inherit one faulty allele, which will be inherited from the infected parent **(1)**

3 *An explanation that makes reference to the following: both polydactyly and cystic fibrosis are caused by gene mutations; Down's syndrome is caused by a chromosome mutation – non-disjunction of chromosome 21 at meiosis; in formation of eggs or sperm; neither parent had Down's syndrome or carried it;
for polydactyly at least one parent has the condition as this is a dominant characteristic;
the child inherits one allele from one parent;
the probability of a child inheriting the condition if one parent is affected is 1 in 2 or 50%;
for cystic fibrosis, which is a recessive characteristic, both parents must be symptomless carriers or one must be affected and one a symptomless carrier;
the child must inherit one faulty allele from each parent **(9)**

111. Dihybrid inheritance

1 AB **(1)**

2 CD; Cd: cD; cd **(4)**

3 (a) genotype of man is cc I^AI^O; genotype of woman is CC I^OI^O **(2)**

(b)

	Father	Mother
parent phenotypes	blood group A with cystic fibrosis	blood group O without cystic fibrosis
parent genotypes	cc I^AI^O	CC I^OI^O
possible gamete genotypes	(cI^A) (cI^O)	(CI^O)

	F (cI^A)	(cI^O)
M (CI^O)	CcI^AI^O	CcI^OI^O

group A without cystic fibrosis and group O without cystic fibrosis (all offspring are heterozygous and are symptomless carriers of cystic fibrosis) **(5)**

(c) 50% / 1 in 2 / a half / 0.50 **(1)**

112. Autosomal linkage

1 C **(1)**

2 B **(1)** *(Statement (3) is not correct as recombinant gametes can form due to crossing over and independent assortment.)*

3 (a) Each chromosome consists of two chromatids; joined at centromere; each has the same gene loci as the original as crossing over has not yet occurred. **(3)**

(b) Recombinant gametes have different combinations of alleles for the linked gene loci, for example, C and d and c and D; formed by crossing over of non-sister chromatids of homologous pairs of chromosomes during prophase I of meiosis. **(2)**

113. Sex linkage and chi squared test

1 (a) it is sex-linked / gene is on X chromosome; males have only one X chromosome so if they inherit a faulty allele they do not have a functioning allele / girls have a functioning allele on their other X chromosome **(2)**

(b) (i) regions of DNA / (premature) mRNA that will be translated into protein **(1)**

(ii) all the genetic material / genes; in a human, body / cells / population **(2)**

(c) Gene has 2 500 000 bases. Exons have 14 000 bases. Introns have been removed, therefore introns contain 2 500 000 − 14 000 bases = 2 486 000 bases. **(2)**

(d) D = gene for dystrophin, d = abnormal allele for non-functioning / absent dystrophin

	Male	Female
parent phenotypes	unaffected (symptomless carrier of DMD)	unaffected (carrier of DMD)
parent genotypes	X^DY	X^DX^d
gamete genotypes	(X^D) (Y)	(X^D) (X^d)
offspring genotypes	X^D X^D X^D X^d	X^D Y X^d Y
offspring phenotype	unaffected female female	unaffected male with male DMD

(5)

114. Selection pressure and genetic drift

1 *Snails are killed and eaten by birds such as thrushes. Predation is a selection pressure. Snails that are better camouflaged will be less likely to be seen by the birds and eaten. They will survive longer and will produce more offspring / young; their offspring may inherit the favourable/advantageous allele(s) making them better camouflaged to avoid predation and survive to reproduce; this happens over many generations and the allele frequency within the population changes; nature selects the snails with the colour that gives best camouflage / blends with the background of their habitat – plain snails blend better against open grassland, and banded snails are better camouflaged in woodlands / hedgerows. **(6)**

2 (a) 10% of 250 = 25 **(1)**

(b) after / due to the typhoon / in 1775, the population went through a genetic bottleneck as it shrank to just 20 people; by chance, at least two of these had the faulty allele; with such a small isolated population there was inbreeding / greater chance of carriers mating and producing offspring; hence greater incidence of the disorder; there is no advantage to being colour blind but it is not disadvantageous enough to cause death so the high incidence has persisted / even if it were very harmful to be colour blind, there is a high incidence of carriers so offspring with the condition will be produced; this is an example of genetic drift **(3)** *(Note: genetic drift can skew allele frequencies in an isolated population but it is not adaptive and does not contribute to evolution in the way that natural selection does.)*

115. Hardy–Weinberg equilibrium

1 C **(1)**

2 no mutations; random mating; large population; no migration; no selection pressures **(5)**

3 The 400 people who cannot smell honeysuckle flowers must be homozygous recessive, so we know that $p^2 = 400 \div 2000$

= 0.2.

$p = \sqrt{0.2} = 0.447$

if $p + q = 1$, then $q = (1 - 0.447) = 0.553$

so $q^2 = 0.306$

The number of people in the 2000 population who are homozygous and can smell honeysuckle is 0.306×2000 = 612

the other equation refers to genotypes: $p^2 + 2pq + q^2 = 1$

So the frequency of heterozygotes is $(1 - (p^2 + q^2))$

= 2000 − (400 + 612)

= 2000 − 1012

= 988 **(4)**

116. Exam skills

1 (a) their chromosome number is halved at meiosis; gametes are haploid; so at fertilisation two (haploid) gametes join to make a diploid zygote **(3)**

(b) during meiosis crossing over during prophase I; alleles are swapped / shuffled, between non-sister chromatids; independent assortment of chromosomes during metaphase I; independent assortment of chromatids at metaphase II; gametes are genetically unique / different from each other and from parent cell; at fertilisation gamete nuclei fuse with unrelated / genetically different gamete nuclei **(5)**

2 (a) (i) one parent is Cc and the other is Cc^h

gametes are ⓒ ⓒ and ⓒ ⓒ^{h}

offspring genotypes CC   Cc^h CC   c^{h}c

offspring phenotypes agouti agouti   agouti   Himalayan

phenotype ratio   3   :   1 (same as 15 : 5) **(1)**

(ii) For the F_1 agouti rat to be mated with a Himalayan rat and produce offspring that are Himalayan and Siamese as well as agouti, the agouti rat must be genotype Cc^h.

As some of the offspring are Siamese, then the Himalayan rat must be genotype cc^h.

parent phenotypes   agouti   Himalayan

parent genotypes   Cc^h cc^h

gametes   ⓒ ⓒ^{h} ⓒ ⓒ^{h}

offspring genotypes Cc   Cc^h c^{h}c   c^{h}c^h

offspring phenotypes agouti agouti   Himalayan   Siamese

phenotype ratio   2   :   1   :   1 **(4)**

(b) mate / breed it with a rat of known genotype, either Siamese (c^{h}c^h) or Himalayan (cc^h); this is a test cross; and observe the phenotypes of the offspring; if any are Himalayan then the agouti male is heterozygous **(4)**

117. Homeostasis

1 **D (1)**

2 Possible answers include:
- Both result in fewer enzyme–substrate complexes being formed and a lower rate of reaction.
- Changing the pH can affect the distribution of (positive and negative) charges in the active site whereas lowering temperature does not.
- Changing the pH can disrupt hydrogen bonds whereas lowering temperature does not.
- Lower temperatures reduce the kinetic energy of the enzyme and substrate whereas changing pH does not.
- Less kinetic energy reduces the number of successful collisions. **(4)**

3 Possible answers include:
- Reduced numbers of ES complexes are formed as temperature increases.
- There is reduced enzyme activity.
- Cellular chemical reactions are negatively affected. **(3)**

118. Hormones in mammals

1 **A (1)**

2 Possible answers include:
- Steroid hormones are lipid based.
- They diffuse through the phospholipid bilayer of the target cell.
- They bind to transcription factors inside the cell. **(2)**

3 Transcription factors are molecules that bind to DNA. Initiate transcription of the anti-sense strand of DNA into mRNA. **(1)**

4 exocytosis; vesicles fuse with the cell surface membrane and hormones are released **(2)**

119. Chemical control in plants

1 **C (1)**

2 when the shoot tip is in the dark there are equal concentrations on either side of the plastic divide; in the light there are 3.8 times as much auxin on the side of the plastic divide furthest from the light; the plant in the dark will grow straight upwards due to the equal distribution; the cells on the shaded side of the plant with higher auxin concentration will become more elongated **(4)**

3 Auxins are involved in apical dominance and prevent lateral bud growth. Removal of the tips would result in no auxin being produced. Without the auxin inhibition, the lateral buds will develop under the stimulation of cytokinins. **(3)**

120. Gibberellin and amylase

1 **D (1)**

2 (a) juglone prevented the germination of the embryoless barley seeds; as no seeds germinated with distilled water and juglone; it reduced α-amylase activity; no α-amylase is being produced by the aleurone layer as embryo is removed; no starch is hydrolysed into maltose / glucose; no glucose production would result in no respiration and no germination **(4)**

(b) Possible answers include:
- some starch hydrolysis will occur
- concentrations of α-amylase will increase over time
- glucose available for respiration **(5)**

121. Phytochrome and photoperiodism

1 (a) leave it in the dark **(1)**

(b) increase in stem length; correct manipulation of the data, e.g. by 23 cm / 18.4%; reference to taller / faster growing seedling; to receive more light / higher red light / to maximise; photosynthesis; idea of allows active phytochrome to be made **(3)**

122. The nervous system

1 **(3)**

Sensory neurone	Motor neurone
cell body in middle of neurone	cell body at end of neurone
short axon	long axon
dendron present	no dendron

2
- motor neurone carries electrical impulses from the CNS
- to an effector such as a muscle or a gland **(1)**

3 Two possible answers from the following:
- correct comparative data to support either statement
- the myelin sheath acts as an insulator
- depolarisation can only occur at the nodes of Ranvier
- saltatory conduction **(5)**

123. The central nervous system

1 (a) **(4)** One mark per correct row.

Region of brain	Name	Function
P	cerebrum	initiates movement
Q	cerebellum	balance
R	medulla	control of heart rate
S	hypothalamus	control of core body temperature

    (b) C **(1)**

2 stimulus – receptor – sensory neurone – relay neurone – motor neurone – effector **(3)** 6 correct = 3 marks; 4/5 correct = 2 marks; 2/3 correct = 1 mark; 0/1 correct = 0 marks

124. Resting potential

1 6250 magnification
   image size = 25 mm = 25 000 μm
   25 000 ÷ 4 = 6 250 magnification **(2)**

2 (a) **(3)**

Description	A	B	C	D	E	F
stage when the concentration of positive ions is greatest inside the axon			✓			
stage when hyperpolarisation first occurs					✓	
stage showing the resting potential	✓					

    (b) sodium–potassium pumps; actively pump potassium ions into the cell; actively pump sodium ions out of the cell **(3)**

125. Action potential

1 D **(1)**
2 (a) (i) **D (1)** (ii) **E (1)** (iii) **B (1)** (iv) **A (1)** (v) **D (1)**
    (vi) **B (1)** (vii) **D (1)**
    (b) if the threshold potential in not reached then there is no action potential; all or nothing law **(2)**

126. Propagation of an action potential

1 positive correlation / as diameter increases the conduction velocity increases; one result at 12 μm diameter does not fit the pattern **(2)**
2 *Three possible answers include:
  • sodium ion can only enter the nerve cell at these nodes of Ranvier / gaps between Schwann cells
  • depolarisation can only occur at these nodes of Ranvier / gaps between Schwann cells
  • action potential jumps from node to node / saltatory conduction **(6)**

127. Synapses

1 A **(1)**
2 Two more possible answers from:
  • depolarisation occurs
  • if threshold is reached an action potential is generated in the postsynaptic nerve cell
  • neurotransmitter is either broken down by enzymes or reabsorbed back into the presynaptic knob **(2)**
3 100 mV **(1)**

128. Drugs and the nervous system

1 C **(1)**
2 Three possible answers include:
  • no / fewer Na^+ channels open
  • no / fewer Na^+ enter cell
  • threshold potential not reached
  • as no depolarisation / not enough depolarisation has occurred **(5)**
3 Four possible answers include:
  • nicotine causes the blood pressure to increase from 130 mm Hg to 195 mm Hg until 6 minutes and then slowly decrease to 150 mm Hg
  • increase of 65 mm Hg
  • binds to nicotinic acetylcholine receptors

  • stimulates the sympathetic nervous system
  • causing vasoconstriction
  • comment that there is a similar effect with the control but there is a larger increase with nicotine
  • increased confidence in conclusion due to small and non-overlapping error bars **(4)**

129. Detection of light by mammals

1 (a) **A (1)**
    (b) correct arrow and **D** (top to bottom) **(1)**
2 *Possible answers include:
  • the sodium channels close and fewer sodium ions enter the rod cell
  • idea of sodium ions pumped out of rod cell
  • hyperpolarisation occurs which stops inhibitory transmitter release at the synapse with the bipolar cell
  • allowing the bipolar neurone to stimulate the sensory neurone to depolarise **(6)**

130. Control of heart rate in mammals

1

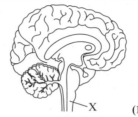

                X   **(1)**

2 Possible answers include:
  • increased muscle contraction and relaxation requires more ATP
  • ATP is produced in respiration
  • more aerobic respiration results in increased carbon dioxide production which diffuses into the blood **(2)**
3 B **(1)**
4 *You are not required to include all of the material indicated below. Additional content included in the response must be scientific and relevant.
  • increase in blood carbon dioxide levels decreases blood pH
  • chemoreceptors detect decrease in blood pH
  • electrical impulse sent to cardiac control centre of medulla
  • increased frequency of impulses sent to sinoatrial node (SAN)
  • involvement of sympathetic nervous system
  • increased frequency of impulses sent to adrenal medulla
  • release of adrenaline into blood
  • heart rate increases
  • decrease in blood carbon dioxide levels increases blood pH back to normal levels
  • chemoreceptors detect increase in blood pH
  • electrical impulse sent to cardiac control centre of medulla
  • increased frequency of impulses sent to SAN via parasympathetic (vagus) nerve
  • heart rate decreases back to normal **(6)**

131. Structure of the mammalian kidney

1 **(4)**

	A	B	C	D	E
Where the glomerulus is located			✓		
Contains podocytes			✓		
Where the collecting ducts are located				✓	
Contains the lowest concentrations of urea		✓			
Affected by ADH				✓	

2 (a)

Amino acid

amine group

One mark for drawing correct amino acid structure and one mark for labelling NH_2 as amine group. **(2)**

(b) ammonia and urea **(2)**

(c) glycogen; and one more from:
- condensation reactions
- 1,4 **and** 1,6 glycosidic bonds formed urea **(3)**

132. Kidney function

1 $(138 \div 775) \times 100 = 17.8\%$ **(2)**

2 The indicative content below is not prescriptive and you are not required to include all the material which is indicated as relevant. Additional content included in the response must be scientific and relevant.

- Ultrafiltration occurs in the glomerulus.
- The filtrate passes into the lumen of the Bowman's capsule.
- Blood cells and large proteins are too large to pass through the gaps in the capillary walls so there are no proteins in the glomerular filtrate.
- All amino acids are actively reabsorbed in the proximal convoluted tubule.
- All glucose molecules are actively reabsorbed in the proximal convoluted tubule.
- Active transport requires energy from ATP.
- So there are no amino acids / glucose molecules in the urine leaving the collecting duct.
- Only some ions are reabsorbed back into the blood.
- Water is reabsorbed back into the blood at several parts of the nephron
- (proximal convoluted tubule / loop of Henle / collecting duct).
- This would contribute to the higher concentrations of urea and inorganic ions in the urine than in the filtrate. **(6)**

133. Osmoregulation

1 C **(1)**

2 (a) collecting duct **(1)**

(b) - ADH causes aquaporins / protein channels to open in the collecting duct
- more water molecules would move out of collecting duct by osmosis
- leading to a more concentrated filtrate **(3)**

3 (a) C **(1)**

(b) Three possible answers include:
- the long loop of Henle produces a lower water potential in the medulla
- counter current multiplier
- more sodium ions would be actively transported out of the ascending limb
- more water molecules would move out of the collecting duct by osmosis **(3)**

134. Thermoregulation

1 **(3)** 3 marks for 5 or 6 correct ticks; 2 marks for 3 or 4 correct; 1 mark for 1 or 2 correct

	Sweating	Vasodilation	Vasoconstriction	Shivering	Contraction of hair erector muscles	Relaxation of hair erector muscles
Increases heat energy loss	✓	✓				✓
Reduces heat energy loss			✓		✓	
Increases heat energy production				✓		

2 *You are not required to include all the material indicated below. Additional content included in the response must be scientific and relevant.
- thermoreceptors detect increase in core body temperature
- electrical impulse generated in sensory neurone
- sent to heat loss centre in the hypothalamus
- autonomic / sympathetic nervous system involvement
- electrical impulses sent via motor neurones
- to effectors
- sweat glands and sphincter muscles in the skin arterioles
- increased sweat production
- evaporation of sweat reduces heat energy in the skin
- heat energy used to break hydrogen bonds between water molecules
- sphincter muscles in skin arterioles relax
- vasodilation
- increased blood flow through skin capillaries
- hair erector muscles relax
- increased heat energy loss through radiation
- negative feedback when core temperature returns to within normal parameters **(6)**

135. Exam skills

1 (a)

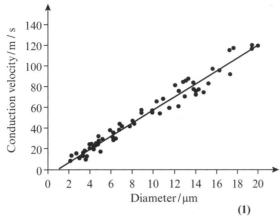

(1)

(b) positive correlation / as the diameter of the neurone increases, so does the speed of the electrical impulse (Correct manipulation of data to support marking point 1.) **(2)**

(c)

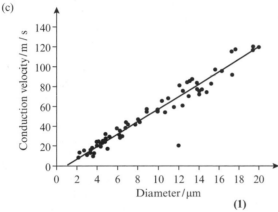

(1)

2 Four possible answers including:
   - animal hormones are produced by glands whereas any plant cell can produce plant hormones
   - animal hormones are transported in the blood, plant hormones diffuse/are transported in xylem to target cells
   - both only affect target cells / cells with complementary receptors
   - both function as chemical messengers
   - both can bring about changes in tissue growth **(4)**

136. Ecosystems

1 habitat; population; community; ecosystem; trophic; primary producers; consumers **(7)**

2 The habitat is the place where an organism lives whereas the niche is its role in the ecosystem. **(1)**

3 *A pyramid of biomass shows the mass of all the organisms collected at one time whereas a pyramid of energy is a more accurate representation as it takes rate of production of biomass and storage of energy into account too. In this ecosystem, the producers reproduce more rapidly than the consumers so even though the biomass and energy storage of producers is less at any one time, it will be more if it is measured across an entire year. **(6)**

137. Ecological techniques

1 A, B, C, D are all correct **(1)**

2 (a) Lay a tape measure up the shore from sea level to the top – this is the transect.
       Place a large quadrat (e.g. 1 m²) at regular intervals along the tape and count the number of periwinkles of each species in the quadrat. **(3)**
   (b) Use a key to identify the different periwinkle species. Repeat the process for three or more belt transects in different positions across the shore.
       Use a statistical test such as the t test to find out whether there are any significant differences in the means for the population sizes. **(2)**
   (c) Repeat in successive years at the same time of year. Draw kite diagrams for the data to show distribution of species across the shore. Kites showing the distribution of a particular species should be drawn one underneath the next for each successive year. **(2)**
   (d) Measurement of abiotic factors, such as temperature, light intensity, wind speed, would increase the amount of information collected. **(1)**

138. Sampling methods

1 (a) Random sampling – use tapes to map area as a grid. Place quadrat randomly and count the number of snails per quadrat. Repeat to ensure data are more accurate. **(2)**
   (b) In order to minimise the time taken to sample while at the same time collecting reliable data, the student should use two quadrats and repeat the sampling seven times. **(2)**
   (c) Spiders are fast moving so traps should be set in random positions across the area. **(1)**

2 C **(1)**

139. Statistical testing

1 (a) $r_s = 0.898$ **(1)**
   (b)   Reject the null hypothesis because r_s is greater than the critical value at $p = 0.05$ so there is a greater than 95% probability that the results are not due to chance and there is a positive correlation between plant height and height on the dune. **(4)**

140. Energy transfer through ecosystems

1 (2)

	Mature rain forest / kJ / m² / yr	Field of alfalfa / kJ / m² / yr	Young pine forest in UK / kJ / m² / yr
GPP	188 000	102 000	51 000
Respiration	134 000	38 000	20 000
NPP	54 000	64 000	3 300

2 54 000 kJ / m² / yr **(1)**

3 $(51\,000 \div 4\,000\,000) \times 100 = 1.3\%$ **(1)**

4 Some light will not fall onto chloroplasts, some will be the wrong wavelength (green light), some will be used to heat the leaves or evaporate water and some will be reflected by the cuticle. **(2)**

5 rate of photosynthesis is higher due to higher temperature; higher light intensity;
   longer days averaged over a year;
   conifers have smaller leaf surface area; limiting absorption of light;
   plant density is lower in pine forest **(4)**

141. Cycling of nutrients

1 B (1)

2 One mark for each vertical column filled in correctly. (4)

Feature of cycle		Nitrogen fixation	Nitrification	Denitrification	Decomposition
The starting point is…	ammonium compounds	✗		✗	✗
	atmospheric nitrogen		✗	✗	✗
	nitrate	✗	✗		✗
	nitrogen-containing organic molecules	✗	✗	✗	
The product is…	ammonia	✗	✗	✗	
	gaseous nitrogen	✗	✗		✗
	nitrates		✗	✗	✗

3 (a) more denitrification in waterlogged soil; so fewer nitrates for plants to convert to amino acids and proteins; no oxygen available for aerobic respiration; so insufficient ATP made for active uptake of nitrates into roots (3)

(b) Waterlogged soils lack oxygen. Oxygen for aerobic respiration is released from nitrate during denitrification. (1)

4 During decomposition, microorganisms use these compounds as energy sources for respiration. Respiration releases CO_2 for photosynthesis. (3)

142. Succession

1 Very few small organisms present at this stage, so do not influence presence of others. However, abiotic conditions are harsh and variable and only few species can survive this. (2)

2 Pioneer organisms complete their life cycle more rapidly, produce many more offspring, tend to be very small and are poor competitors. Equilibrium species are larger, better competitors and slower to complete their life cycle, producing fewer offspring. (2)

3 (a) Humus increases the nutrient content of the soil and releases, for example, nitrates and phosphates as it is decomposed. It also increases moisture retention in the soil. (2)

(b) Primary productivity depends on the amount of photosynthesis. As succession proceeds, primary productivity increases because there are a higher number of larger plants. As succession continues, the primary productivity will eventually level off because the climax community is at equilibrium and there is no further increase in plant biomass. (4)

143. Biotic and abiotic factors

1 (a) at high tide the levels return to those of the sea (1)

(b) no change as rock pool remains covered (1)

(c) Algae are photosynthesising faster than they are respiring. Photosynthesis produces oxygen and so oxygen builds up. (1)

(d) CO_2 levels fall in the pool as it is used up during photosynthesis. As CO_2 is an acid gas, the pH rises as the CO_2 level drops. (1)

(e) Increase in O_2 means more available for aerobic respiration. Increase in pH will affect enzyme function and reduce reaction rates. (2)

2 Sundew catches and digests insects as a source of nutrients, such as nitrate and phosphate, so is able to grow in the nutrient poor soil. Roots have little oxygen available for aerobic respiration as soils are waterlogged, but as they will not need energy for active uptake of nutrients from the soil this will not matter. Sundew is not found in areas where other plants grow as it is grows slowly and therefore is unable to compete. (3)

144. Abiotic factors and morphology

1 (a) This takes the size and the overall shape of the shell into account, making it possible to compare different shells. (1)

(b) *Hypothesis: The shells of the limpets on the upper shore have a different height to width ratio from those on the lower shore.

Independent variable: Height and width of limpet shells measured using Vernier callipers.

Dependent variable: Position on shore – upper and lower shore identified by seaweed zonation.

Controlled variables: Position on shore, species of limpet.

Uncontrolled variables: Many factors including type of rock and weather conditions. The two sites are in the same location so these should be similar.

Valid results: Limpets sampled randomly on each area of shore. Readings repeated 20 times and means calculated. (6)

(c) Student's t test would be best as it allows means to be compared statistically. Alternatively, the standard deviation of each mean could be calculated and used to plot error bars on a graph of the results – if the error bars overlap then the difference is not significant. (2)

(d) Limpets suffer from desiccation when the tide goes down. The tide is out for longer higher up the shore so there imore desiccation. Narrower width means less water loss from the shell aperture. (1)

(e) The limpets which are higher relative to their total size will be more likely to be knocked off the rocks by waves. (1)

145. Human factors and climate change

1 D (1)

2 D (1)

3 When solar radiation (UV) enters the atmosphere, it is re-radiated at a longer wavelength (IR) at a lower energy level. Greenhouse gases absorb this lower energy infrared radiation re-radiated from the Earth's surface and prevent heat from escaping out of the atmosphere. This results in higher global temperatures. (3)

4 It maintains global temperatures at a level suitable for life – neither too hot nor too cold; prevents extremes. (1)

5 many extinctions around this term – disappearance of fossil types (1)

6 *You are not required to include all the material indicated below. Additional content included in the response must be scientific and relevant.

- fossil fuels produce CO_2
- no direct evidence that higher CO_2 means global warming but data from 1900 shows positive correlation between CO_2 levels and global temperatures, long-term data show the same pattern
- CO_2 also produced during decomposition
- active volcanoes also produce CO_2
- removing carbon sinks increases CO_2 production
- methane and other greenhouse gases produced by paddy fields
- acidification of oceans releases CO_2
- warming may be part of natural cycle
- data from the past used to show this
- scientists may be biased as they may have vested interests

(6)

146. Human factors and sustainability

1 extinction rates much higher than previously **(1)**
2 Conservation involves active management with the aim of sustainability. Preservation simply means maintaining the status quo. **(1)**
3 Indigenous people have no choice in how they live and rely on resources for their own survival. They will also have an interest in ensuring that resources are maintained at a sustainable level. **(1)**
4 Convention on International Trade in Endangered Species; international agreement limiting or preventing import or export of endangered species and their products **(2)**
5 *You are not required to include all the material indicated below. Additional content included in the response must be scientific and relevant.
   Protected reserves:
   - animals can continue to exhibit natural behaviour – less stress
   - no need to feed them as natural food available
   - easy to release them back into wild afterwards
   - can continue to play a part in food webs
   - may be difficult to protect from poachers
   - if reserve is large, animals not easy to monitor (*max 3 marks*)
   Captive breeding programmes:
   - animal health can be carefully monitored and disease treated rapidly
   - breeding is controlled so decisions can be made about which animals should be mated to maintain maximum genetic diversity
   - may not be possible for animals to be released back into wild as they may have lost the necessary skills. (*max 3 marks*) **(6)**

147. Exam skills

1 (a) GPP is gross primary productivity and is the total amount of energy in organic molecules trapped by photosynthesis per unit area per unit time. **(2)**
   (b) NPP = 4680; R = 5720 **(2)**
   (c) NPP = GPP − R; 55% energy in GPP is lost; as heat; during life processes, e.g. active transport, DNA replication, etc; NPP is energy available for next trophic level; NPP can be optimised by growing young plants which convert proportionately more energy into biomass **(3)**
   (d) cattle are primary consumers so gain energy from NPP; farmer needs to ensure there is sufficient NPP available for all cattle; or yield will be reduced **(3)**
   (e) gives an average value across the whole year as the biomass is likely to change depending on the season **(2)**
   (f) this grass has a higher rate of photosynthesis; lower rate of respiration; more nutrients in soil **(2)**

TIMED TEST ANSWERS

AS Level timed test 1: Core Cellular Biology and Microbiology

Additional guidance on mark allocation has been provided in some cases.

1 (a) hydrolysis **(1)**
   (b) • reactants: sucrose and water
      • products: glucose and fructose **(2)**
   (c) • Way in which conditions might be controlled: keep temp low / keep pH sub-optimal / keep enzyme concentration low.
      • Explanation: manipulate the conditions so the enzyme cannot work at its optimum rate – there is a limiting factor. **(2)**
2 (a) (i) **C (1)**
      (ii) **B (1)**
   (b) An explanation that makes reference to some of the following:
      • DNA unzips because hydrogen bonds are broken using helicase
      • mRNA made by complementary base pairing / transcription / using RNA polymerase
      • mRNA leaves nucleus and attaches to ribosomes / rRNA
      • tRNA anticodon attaches to mRNA codon
      • peptide bonds formed between amino acids on the ribosome as mRNA is translated into polypeptide chain **(4)**
3 (a)

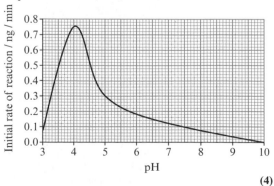

   (2)

(b) An explanation that makes reference to some of the following:
   • straight chain with β1,4 glycosidic links
   • each cross linked to others with hydrogen bonds
   • to form microfibrils
   • giving strong multi-fibre structure (like rope)
   • β1,4 glycosidic links not easily hydrolysed
   • so fibres last a long time
   • arrangement of microfibrils gives an open structure to allow materials in and out of cell / ensures wall permeable **(3)**
   Maximum of 3 marks for structural points.
(c) Calculation of necessary rates: mark for explanation of how to do this as well as for the correct maths
   • pH 3 = 0.3 ÷ 4 = 0.075
   • pH 4 = 3 ÷ 4 = 0.75
   • pH 5 = 1.2 ÷ 4 = 0.3
   • pH 7 = 0.5 ÷ 4 = 0.125
   • pH 10, zero
   all five = 2; one wrong = 1; two wrong = 0
   • *y*-axis scale appropriate
   • plots accurate

[Graph: Initial rate of reaction / ng / min (y-axis, 0.0 to 0.8) vs pH (x-axis, 3 to 10). Curve rises sharply to a peak around pH 4 (~0.75) then decreases gradually towards zero at pH 10.]

   (4)

4 (a) 32% guanine, this means 32% also cytosine = 64%
   • thus 36% A and U
   • so 18% U which is 0.18 × 450 = 81 uracil **(3)**

(b)

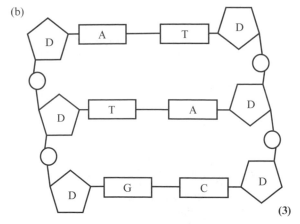

(3)

(c) An explanation that makes reference to some of the following:
- the pairing referred to is between bases
- the specific aspect is that adenine (A) always pairs with thymine (T) and guanine (G) always pairs with cytosine (C)
- the copying mechanism is the unzipping of the DNA to expose unpaired bases
- which can then pair with bases on free nucleotide to form two new strands **(4)**

Additional marking point: because the pairing is specific, the new strands will always be identical to the original strands.

5 (a) *An explanation that makes reference to some of the following:
- sperm cell fuses with egg cell (membrane)
- reference to cortical granules / vesicles / lysosomes
- idea of (cortical granules) moving towards / fusing with egg cell (surface) membrane
- from the secondary oocyte
- reference to exocytosis (of cortical granules / vesicles / lysosomes)
- idea of contents (of cortical granules) secreted / released into jelly layer / OR reference to cortical reaction
- idea of hardening / thickening / of zona pellucida / jelly layer OR formation of fertilisation membrane
- reference to change in charge across egg cell membrane **(6)**

(b) (i) • run in water bath / incubator at set temperature / stated temperature 15–35 °C as temperature can affect pollen tube growth
- ensure pH is always the same, check and use buffer if necessary, because pH can affect pollen tube growth
- make sure pollen grains all from same plant / flower / genetically identical, as pollen tube growth might have aspects which are genetically controlled **(3)**

(ii) • up to 10% sucrose, an increase in sucrose increases (mean) length of pollen tube / positive correlation; greatest increase between 5% and 10%
- greatest (mean length of pollen tube) at 10%
- idea that above 10% the pollen tubes are shorter, e.g. negative effect or correlation
- credit correct manipulation of the data, e.g. 570–580 µm longer when grown in 10% sucrose compared to 0% sucrose
- appropriate comment on significance of overlapping error / range bars between 5% and 30% / 10% and 20% **(3)**

6 (a) (i) orcein **(1)**
(ii) **B (1)**

(iii)

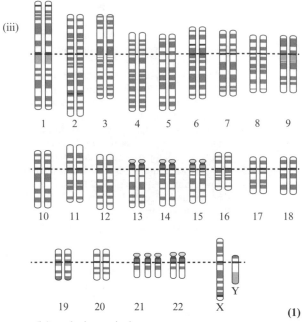

(1)

(iv) • during meiosis
- when pairs of homologous chromosomes should separate, pair 21 fails to do so
- this gives rise to a gamete with two copies of 21
- when this joins with a normal gamete it gives rise to a zygote with three copies of chr. 21 **(3)**

(b) as age increases so does incidence
- from 25 to 35 increase is only 2.2 ÷ 1000 but from 35 to 45 it is 17 ÷ 1000
- between 40 and 50 increase is 400%
- huge increase between 45 and 50 **(4)**

7 (a) (i) Archaea are as different from Bacteria as humans are different from Bacteria so Woese decided there must be three major groupings. **(1)**
(ii) • peer review
- conferences
- experiments / etc. repeated by others **(3)**

(b) (i) they are non-living **(1)**
(ii) reverse transcriptase inhibitors, B
attachment inhibitors, A
protease inhibitors, C
integrase inhibitors, D **(2)**
(iii) • may kill the host before they have a chance to infect other people
- newly evolved strain so it has not yet become reduced in virulence
- reference to isolation of victims **(3)**

(c) • both designed to replicate viruses
- both involve virus genes getting into host
- in lytic, the virus genes take over host synthetic machinery; in lysogenic, enter host DNA
- in lytic new viruses made straight away; in lysogenic may be long wait
- in lysogenic the viral DNA is replicated with the hosts
- this does not happen in lytic **(4)**

8 (a) (i) 0.3 ÷ 50;
= 0.006;
arbitary units / minute **(3)**
(ii) axes correct and labelled (x-axis = enzyme conc. / AU and y-axis = initial rate / absorbance / 1000 s
- all plots correct
- line dot to dot or best fit **(3)**
(iii) as enzyme concentration rises there are more active sites;
therefore more substrate can be converted in unit time;
at 70 AU substrate becomes limiting so rate does not rise any further **(3)**

(b) • in the rough endoplasmic reticulum (RER) the protein is folded
- and it is then packaged into (transport) vesicles by the RER
- these vesicles fuse to form the Golgi apparatus
- the protein is changed in the Golgi apparatus
- and then transferred inside secretory vesicles from the Golgi apparatus to the cell (surface) membrane
- these vesicles fuse with cell membrane (exocytosis) **(5)**

AS Level timed test 2: Core Physiology and Ecology

1 (a) **C (1)**
   (b) **C (1)**
   (c) **A (1)**
   (d) (i) choose any time on 2 May and same time on 6 May, e.g. the highest point on 2 May and the highest on 6 May;
   $= 4320 - 4271 = 49 \ \mu$m;
   so this is $49 \div 96 \ \mu$m / h $= 0.51 \ \mu$m / h;
   $= 0.00051$ mm / h;
   which is 5.1×10^{-4} mm / h **(5)**
   (ii) An explanation which makes reference to the following:
   - xylem vessels make up the wood of the stem
   - during the day they are under tension (suction) due to rapid transpiration
   - this reduces the diameter of each one and thus the whole twig
   - in darkness the transpiration rate falls as stomata close (less photosynthesis)
   - xylem vessels under much less tension as water flow slows and the twig diameter increases **(4)**
   (e) • xylem is under tension (suction) but phloem is under positive pressure
   - therefore it would be predicted that the muscle of froghoppers would be bigger than that of aphids
   - as they will need to exert a much greater suction force on the xylem sap **(3)**

2 (a) • both involve swelling of a body part due to accumulation of tissue fluid
   - kwashiorkor swelling of the belly; filariasis swelling of the limbs
   - kwashiorkor caused by dietary deficiency; filariasis caused by parasitic worms
   - swelling in kwashiorkor caused by lack of protein affecting oncotic pressure of blood and so reabsorption of tissue fluid into lymph vessels; filariasis caused by blocking of lymph vessels by worms so the lymph cannot drain **(4)**
   (b) (i) State all six of the following:
   - Contact between platelets and damaged tissues makes them break open releasing clotting factors including thromboplastin.
   - Thromboplastin is an enzyme that converts plasma protein prothrombin to its active form, thrombin,
   - in the presence of calcium ions.
   - Thrombin is an enzyme that acts on the soluble protein fibrinogen.
   - It converts it into insoluble fibrin that forms a mesh of fibres.
   - Platelets and blood cells get caught in the fibrin mesh forming a clot. **(5)**
   (ii) An explanation which includes the following:
   - damage to artery wall
   - white blood cells move into wall (of artery) / inflammatory response
   - cholesterol deposited / reference to foam cells
   - fibrous tissue develops / reference to plaque / atheroma; artery wall hardens / becomes less elastic / atherosclerosis
   - clot forms due to damage to plaque or further damage to artery wall – blocks vessel to brain – stroke **(4)**

3 (a) • humans more closely related to chimp (than to orang-utan and gorilla)
   - reference to humans and chimps more closely related to orang-utan than gorilla
   - reference to similarity of sequence indicates closeness of ancestral relationship
   - human and chimp sequence identical
   - orang-utan has one difference, gorilla has two differences
   - reference to number 19 for orang-utan / number 9 and 19 for gorilla
   - *must have* for one mark: data shows how recently the different apes split away from a common ancestor, e.g. all had common ancestor, then gorillas split away, then orang-utans, then chimps and humans **(4)**
   (b) (i) A description which includes some of the following:
   - DNA split into fragments
   - using restriction endonucleases.
   - DNA fragments are added to a gel
   - containing a dye that fluoresces under ultraviolet (UV) light.
   - A current is passed through the apparatus.
   - The DNA fragments move towards the positive anode.
   - The plate is placed under UV light.
   - The DNA fluoresces and shows as a pattern of bands. **(5)**
   (ii) • genes are the codes for protein
   - genes are sections of DNA
   - so similarity (of DNA) indicates closeness of relationship **(2)**

4 (a) (i) • both have decreased
   - decrease in roadside verges greater than in hedgerows
   - percentage / proportional decrease greater in roadside verges than in hedgerows
   - roadside verges greater species richness at beginning and end of decade **(4)**
   (ii) • biodiversity is more than species richness
   - species richness gives no indication of abundance of each species
   - use of a calculated biodiversity index, e.g. Simpson's, gives much better guide
   - species richness gives no indication to genetic diversity **(4)**
   (b) • reference to seeds stored in cool, dry conditions
   - seeds can be stored for long time
   - viability tests carried out at regular intervals
   - more economic / less costly / less labour involved than conserving living plants
   - less space needed
   - large numbers of plants can be stored
   - seeds do not need be stored in original habitat
   - less likely to be damaged by vandalism / natural disaster / disease / herbivores
   - not all types of plants produce seeds that can be stored
   - a major power cut or natural disaster could wipe out all the reserves as can store many species in a single site **(5)**

5 (a) (i) idea of secretion of waxy substance **(1)**
   (ii) • active at night / inactive in day
   OR
   - spreads wax over skin **(1)**
   (iii) hunts in trees rather than on the ground **(1)**
   (b) • it eats insects at night in trees
   - within the community / ecosystem / habitat / environment / in hot, dry areas with trees **(2)**

(c) • selection pressure / change in environment / hot and dry habitat
  • reference to competition / predation
  • mutation (in frog)
  • advantageous allele arises, e.g. allele for waxy secretions
  • individuals with advantageous alleles / characteristics can survive and breed
  • advantageous allele / mutation passed on (to future generations)
  • leading to increased frequency of advantageous alleles in the population **(5)**

6 (a) • (fenitrothion) increases the permeability of membranes
  • cell surface membrane is affected before vacuole membrane, shown by mineral ions leaking out immediately because only the cell surface membrane needs to be damaged
  • for betalain leakage both cell surface membrane and the vacuole membrane would need to be damaged so there is a delay before the increased leakage is seen **(4)**

(b) Answers will be credited according to the candidate's deployment of knowledge and understanding of the material in relation to the qualities and skills outlined in the generic mark scheme.
The indicative content below is not prescriptive and candidates are not required to include all the material which is indicated as relevant. Additional content included in the response must be scientific and relevant.
Indicative content:
  • reference to 3 / more different pH values used as this is the independent variable
  • validity ensured by detailing how pH varied, e.g. buffer
  • repeats at each pH to estimate reliability of data
  • obtain beetroot samples from same root
  • control of other named variable to ensure validity, e.g. temperature, concentration of fenitrothion, size of discs
  • safety aspect explained, e.g. cut finger when cutting beetroot + cut away from finger / cut on a hard surface **(6)**

7 (a) *Any three examples from the following, e.g.
  • left and right sides separate / septum keeping the oxygenated and deoxygenated blood apart
  • reference to muscular nature of walls of ventricles compared to atria for their function of pumping blood out of the heart rather than between chambers
  • reference to cardiac muscle – does not fatigue / own blood supply through coronary arteries
  • relative thickness of ventricle (walls) – right ventricle pumps blood to lungs – relatively low pressure as delicate tissue of lungs and relatively short distance to go, left pumps against resistance of arterial system and has to reach all over the body
  • there are other points – if structural adaptation highlighted and linked correctly to function, give 2 marks **(6)**

(b) A description which makes reference to the following:
  • intrinsic rhythmicity – required for full marks
  • (Wave of electrical) impulses / depolarisation from node
  • passes over (both) atria sinoatrial
  • resulting in atrial systole / contraction.
  • slight delay at the atrioventraicular node.
  • (Impulses) pass along bundle of His
  • (up / along) Purkyne fibres.
  • Correct direction of impulse described / ventricles contract from the base up
  • resulting in ventricular systole / contraction. **(4)**

A Level timed test 1: Advanced Biochemistry, Microbiology and Genetics

1 (a) • correct glycosidic bond
  • correct β-glucose molecules

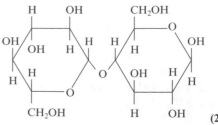

(2)

(b) glycosidic **(1)**

(c) An explanation that makes reference to some of the following (*maximum of 3 marks for structural points*)
  • straight chain with multiple β1,4 glycosidic links
  • each cross linked to others with hydrogen bonds
  • to form microfibrils
  • giving strong multi-fibre structure (like rope)
  • β1,4 glycosidic links not easily hydrolysed
  • so fibres last a long time
  • arrangement of microfibrils (gives an open structure to allow materials in and out of cell /ensures wall permeable) **(4)**

(d) D **(1)**

2 (a) the complete set of (genes / genetic material) present in a cell / organism **(1)**

(b) An answer to include two of the following similarities:
  • DNA polymerase
  • DNA nucleotides with bases A, C, G, T
An answer to include two of the following differences:
  • DNA replication uses helicase to separate the strands, PCR heats the strands to 90 °C to separate the strands
  • copies the whole of the genome once, PCR makes many copies of small fragments of DNA **(4)**

3 (a) C **(1)**

(b) *Answers will be credited according to candidate's knowledge and understanding of the material in relation to the qualities and skills outlined in the generic mark scheme. The indicative content below is not prescriptive and candidates are not required to include all the material which is indicated as relevant. Additional content included in the response must be scientific and relevant.
  • vaccination contains smallpox (non-self) antigens
  • phagocytosis
  • phagocytes / macrophages become antigen presenting cells (APCs)
  • APCs activate T helper cell with complementary shaped receptor
  • clonal selection of B lymphocyte / T killer cell
  • differentiation into memory B cell and B effector / plasma cells
  • production of antibodies
  • differentiation into T killer memory cells and active T killer cells
  • idea that memory cells ensure faster response if infected with same non-self antigen so pathogen destroyed before symptoms occur **(6)**

(c) (i) A **(1)**
  (ii) • line increases before day 5
    • steeper increase than vaccination line
    • higher concentration of antibodies than vaccination line **(3)**
  (iii) An explanation which makes reference to the following:
    • antibodies would be produced faster
    • as clonal selection would occur quicker
    • antibodies would be produced in greater numbers / quantities
    • as there are larger numbers of memory / plasma cells

Answers must be comparative. **(4)**

4 (a) (i) A description to include the following:
- restriction endonuclease used to cut bacterial DNA
- produce sticky ends complementary to the clotting factor gene sequence
- DNA ligase causes phosphodiester bonds to join DNA together **(3)**

(ii) An answer to include one of the following:
- bacteriophage / virus which infects bacteria
- gene gun **(1)**

(b) A description to include four of the following:
- antibiotic resistance marker gene also inserted
- culture bacteria on medium containing bactericidal antibiotic
- suitable incubation temperature / aerobic conditions
- only recombinant bacteria will survive
- correct reference to replica plating **(4)**

(c) (i) **B (1)**

(ii) An answer which makes reference to the following:
- $1\frac{1}{4} \div 1.25$ hours **(1)**

(iii) Example answer as candidates may use other points to calculate gradient:
- (540-300)/8
- 30 **(2)**

(iv) An answer which makes reference to the following:
- B has a positive gradient as the population is increasing / whereas D has a negative gradient as the population is decreasing / more *E.coli* are dying than dividing **(1)**

5 (a) • correct readings
- correct answer **(2)**

(b) (i) **A (1)**

(ii) **C (1)**

(c) (i) • correct readings
- correct answer **(2)**
(11 ÷ 200) × 100 = 5.5%

(ii) An answer which includes the following:
- slide B
- meristem tissue is found at the root tip
- no cells are undergoing mitosis in slide B so cannot be meristematic tissue **(3)**

(d) An explanation which includes the following:
different from other sperm cells:
- random assortment of chromosomes during meiosis
- crossing over results in recombination of alleles in different ways / chromatids / chromosomes containing different alleles
different from body cells:
- haploid not diploid **(3)**

6 (a) (i) **B (1)**

(ii) **D (1)**

(b) An answer which includes three of the following:
- hydrogen ions are pumped out of the companion cell
- using ATP
- sucrose is loaded by active transport
- sucrose diffuses down the concentration gradient into the sieve tube element **(3)**

(c) An explanation which includes three of the following:
- fructose is soluble whereas fructans are insoluble
- fructose would affect / lower the water potential of a cell, causing water to enter
- by osmosis
- fructans do not affect the water potential of the cell **(3)**

(d) *Answers will be credited according to candidate's knowledge and understanding of the material in relation to the qualities and skills outlined in the generic mark scheme.
The indicative content below is not prescriptive and candidates are not required to include all the material which is indicated as relevant. Additional content included in the response must be scientific and relevant.*
- reference to suitable number of germinating and non-germinating seeds
- Source of cell material with same surface area / mass / source / age / solutions with controlled pH / volume, as these factors can affect the validity of the results obtained
- washed in sterile water / reference to appropriate precautions taken to minimise risk
- weigh seeds before experiment
- seeds place in solutions of different molarities of sugar / glucose solution
- reference for suitable time
- blotted dry and reweighed
- or percentage of plasmolysed cells counted using microscope
- reference to control of temperature
- reference to replicates/repeats
- ref to drawing graph
- ref to comparing water potential of germinating and non-germinating seeds
- ref to statistical analysis **(6)**

7 (a) (i) **D (1)**

(ii) An explanation which includes five of the following:
- both are globular proteins
- haemoglobin has a quaternary structure whereas most enzymes have a tertiary structure
- haemoglobin has a haem group / iron ion whereas enzymes do not
- haemoglobin has no active site, enzymes do
- both have amino acids as monomers
- hydrogen / ionic bonds present
- α helix / β-pleated sheet **(5)**
Answers must be comparative.

(b) (i) 4 **(1)**

(ii) An answer which includes the following:
- A
- fetal haemoglobin has a higher affinity for oxygen
- leading to higher percentage saturation than adult haemoglobin (at the same partial pressures of oxygen) **(3)**

(c) One mark per correct row
- competitive inhibition is where non-substrate molecule fits into the enzyme active site / non-competitive inhibitors fit into allosteric site / bind to a site other than the active site
- competitive inhibition blocks the active site / non-competitive inhibitors change the shape of the active site
- both reduce the rate of reaction **(3)**

8 (a) • correct readings
- correct answer **(2)**
0.8 × 60 = 48 bpm

(b) **C (1)**

(c) A description which includes the following:
- electrical impulses produced by sinoatrial node
- electrical impulses travel from atrioventricular node down the bundle of His / Purkyne fibres
- ventricular systole / contraction of muscular walls of ventricle **(3)**

(d) An explanation which includes the following:
- thinner smooth muscle layer as less strength needed to withstand pressure
- thinner elastic layer as vein will not need to recoil to original shape
- larger lumen to provide less resistance to blood flow
- valves to prevent backflow of blood **(4)**

A Level timed test 2: Advanced Physiology, Evolution and Ecology

Additional guidance on mark allocation has been provided in some cases.

1 (a) (i) A description that makes reference to the following:
- idea of peer review
- idea of repeating observations to confirm or validate findings **(2)**

(ii) **C (1)**
(b) **D (1)**
(c) **A (1)**
(d) **A (1)**
(e) (i) 4.5–2.5
  80% **(2)**
  (ii) An explanation which makes reference to the following:
  • correct ref to diffusion (of substance B) occurring due to concentration difference
  • idea of rate of uptake decreases as the concentration gradient decreases
  • (net) uptake stops when concentration inside cell equals that outside the cell **(3)**
  (iii) An answer which makes reference to two of the following:
  • active transport is against concentration gradient whereas diffusion is not
  • active transport requires ATP whereas diffusion is passive
  • reference to involvement of (membrane) proteins in active transport whereas they are not involved in diffusion **(2)**
  Answers must be comparative.

2 (a) (i) **D (1)**
  (ii) **D (1)**
  (iii) **A (1)**
  (b) An explanation that includes four of the following:
  • light absorbed by rhodopsin
  • rhodopsin changes shape
  • rhodopsin is converted to retinal AND opsin
  • opsin binds with cell surface membrane
  • idea of fewer sodium ions / Na^+ enter rod cell / sodium ions pumped out of rod cell
  • hyperpolarisation occurs (leading to change in voltage) **(4)**
  (c) An explanation to include the following:
  • iodopsin is less sensitive to light than rhodopsin
  • a single cone cell synapses with a single bipolar cell **(2)**
  Accept reverse argument.

3 (a) the feeding positions in a food chain or web **(1)**
  (b) • correct readings
  • correct answer **(2)**
  $2800 - 1750 = 1050$
  $(1050 \div 5300) \times 100 = 19.8\%$
  (c) All rows correct = 3 marks
  4 or 5 rows correct = 2 marks
  2 or 3 rows correct = 1 mark

	Trophic level 1	Trophic level 2	Trophic level 3	Trophic level 4
Autotroph	✓			
Carnivore			✓	✓
Herbivore		✓		
Heterotroph		✓	✓	✓
Primary consumer		✓		
Tertiary consumer				✓

(3)

  (d) (i) An explanation which includes reference to the following:
  • production of / energy incorporated into biomass / organic material / organic molecules / tissue in producers / plants
  • reference to losses in respiration / GPP-R **(2)**
  (ii) An answer which includes reference to four of the following:
  • NPP depends on photosynthesis
  • higher the temperature the more NPP
  • enzymes in photosynthesis can work faster
  • increase in rainfall increases NPP
  • water needed for light-dependent reaction
  • role of water in transport **(4)**

4 (a) (i) **B (1)**
  (ii) **A (1)**
  (b)

Point on diagram	Permeable to potassium ions	Permeable to sodium ions
P		✓
Q	✓	

(2)

  (c) (i) *Physostigma* **(1)**
  (ii) Answers will be credited according to candidate's knowledge and understanding of the material in relation to the qualities and skills outlined in the generic mark scheme.
  The indicative content below is not prescriptive and candidates are not required to include all the material which is indicated as relevant. Additional content included in the response must be scientific and relevant.
  • extra indicative content =
  • - ref to action of inhibitors
  • - acetylcholinesterase can't form enzyme-substrate complexes
  • acetylcholine would not be broken down
  • would remain in the synaptic cleft / would keep binding to postsynaptic receptors
  • postsynaptic Na^+ channels open and sodium ions enter the postsynaptic cell
  • causing the postsynaptic membrane to depolarise.
  • correct reference to excitatory postsynaptic potentials **(6)**

5 (a) An explanation which includes the following:
  • correct reference to obtaining DNA sample
  • use of restriction enzymes to cut DNA
  • correct reference to gel electrophoresis
  • correct reference to counting the number of genes with heterozygosity / see how many bands match **(4)**
  (b) An explanation which includes two of the following:
  • fewer alleles in gene pool which could confer selective advantage when environment changes
  • correct reference to lack of adaptation / example / no selective advantage (when environment changes)
  • (therefore) less likely to survive
  • (therefore) more at risk of extinction **(2)**
  (c) An answer which includes two of the following:
  • greater genetic diversity amongst the cubs
  • greater genetic fitness
  • greater chance that some cubs will survive and reproduce
  • increase in population size **(2)**
  (d) An explanation which includes the following:
  • idea that genetic diversity (GD) considers one species but species richness (SR) considers different / number species
  • idea that GD considers alleles / genotypes but SR is within a habitat / area **(2)**
  (e) • 2 (0.3)(0.7)
  • 42
  • % **(3)**

6 (a) An explanation which makes reference to the following:
  • (receptor / transcription factor) binds to promoter
  • stimulates RNA polymerase / RNA polymerase binds
  • transcription of gene **(3)**
  (b) other cells do not have the oestrogen receptors **(1)**
  (c) **D (1)**
  (d) (i) $(180 \div 100) \times 7.5\% = 13.5$ **(2)**
  (ii) An answer which makes reference to the following:
  • ADH causes channels / aquaporins to open in collecting duct / makes collecting duct permeable
  • medulla has lower water potential
  • water moves by osmosis into medulla **(3)**

7 (a) A description which includes three of the following:
  • line transect
  • quadrats placed every 0.5 m

- measurement of % density of both species in each quadrat
- replicates to allow a mean density at each distance to be calculated **(3)**

(b) A description which includes three of the following:
- species A has a lower density at 0 m from the road than species B
- the mean density of species B decreases with increasing distance from the road / no species B is found further than 2 m from the road
- mean density of species A fluctuates, but is most common at 1 m from the road
- correct manipulation of data to reinforce any point above **(3)**

(c) *Answers will be credited according to candidate's knowledge and understanding of the material in relation to the qualities and skills outlined in the generic mark scheme.

The indicative content below is not prescriptive and candidates are not required to include all the material which is indicated as relevant. Additional content included in the response must be scientific and relevant.
- reference to suitable number of plants
- source of plants with same age / solutions with controlled pH / volume, as these factors can affect the validity of the results obtained
- different molarities / concentration of salt solution for watering
- reference for suitable time
- height of plant measured at start
- height of plant measured at end
- reference to control of temperature / light intensity
- reference to replicates / repeats
- reference to drawing graph
- reference to comparing growth against salt concentration
- reference to statistical analysis to see if there is a significant difference **(6)**

Published by Pearson Education Limited, 80 Strand, London, WC2R 0RL.

www.pearsonschoolsandfecolleges.co.uk

Copies of official specifications for all Edexcel qualifications may be found on the website: www.edexcel.com

Text © Pearson Education Limited 2016

Typeset and illustrations by Tech-Set Ltd, Gateshead
Produced by Out of House Publishing
Cover illustration by Miriam Sturdee

The rights of Gary Skinner, Ann Skinner, Deborah Eldridge, Sue Hocking, Caroline Radbourne-Harris and Hilary Otter to be identified as authors of this work have been asserted by them in accordance with the Copyright, Designs and Patents Act 1988.

First published 2016

19 18 17 16
10 9 8 7 6 5 4 3 2 1

British Library Cataloguing in Publication Data
A catalogue record for this book is available from the British Library

ISBN 978 1 447 98993 6

Printed in Slovakia by Neografia

Acknowledgements
We are grateful to the following for permission to reproduce copyright material:

Quote Timed Tests.4c from Molecular Structure of Nucleic Acids: A Structure for Deoxyribose Nucleic Acid, Nature 171, pp.737-738 (J. D. Watson and F. H. C. Crick 1953), Copyright © 1953, Rights Managed by Nature Publishing Group, Reprinted by permission from Macmillan Publishers Ltd

The publisher would like to thank the following for their kind permission to reproduce their photographs:

(Key: b-bottom; c-centre; l-left; r-right; t-top)
Alamy Images: blickwinkel 33; **Dr. Leighton Dann:** 39; **Gary Skinner:** 53; **Professor Ruiqin Zhong:** 27; **Science Photo Library Ltd:** John Durham 26, John Greiim 155r, Mauro Fermariello 155l, Prof S. Cinti 28, Sinclair Stammers 160, Steve Gschmeissner 124, Thomas Deerinck, NCMIR 50; **Shutterstock.com:** Aleksey Stemmer 156, Kathathep 121

All other images © Pearson Education

A note from the publisher
In order to ensure that this resource offers high-quality support for the associated Pearson qualification, it has been through a review process by the awarding body. This process confirms that this resource fully covers the teaching and learning content of the specification or part of a specification at which it is aimed. It also confirms that it demonstrates an appropriate balance between the development of subject skills, knowledge and understanding, in addition to preparation for assessment.

Endorsement does not cover any guidance on assessment activities or processes (e.g. practice questions or advice on how to answer assessment questions), included in the resource nor does it prescribe any particular approach to the teaching or delivery of a related course.

While the publishers have made every attempt to ensure that advice on the qualification and its assessment is accurate, the official specification and associated assessment guidance materials are the only authoritative source of information and should always be referred to for definitive guidance.

Pearson examiners have not contributed to any sections in this resource relevant to examination papers for which they have responsibility.

Examiners will not use endorsed resources as a source of material for any assessment set by Pearson.

Endorsement of a resource does not mean that the resource is required to achieve this Pearson qualification, nor does it mean that it is the only suitable material available to support the qualification, and any resource lists produced by the awarding body shall include this and other appropriate resources.

NOTES

NOTES